MOUNTAINS OF FIRE

MOUNTAINS OF FIRE

THE MENACE, MEANING, AND MAGIC OF VOLCANOES

CLIVE OPPENHEIMER

The University of Chicago Press

The University of Chicago Press, Chicago 60637
© 2023 by Clive Oppenheimer
All rights reserved. No part of this book may be used or reproduced in any
manner whatsoever without written permission, except in the case of brief
quotations in critical articles and reviews. For more information, contact
the University of Chicago Press, 1427 E. 60th St., Chicago, IL 60637.
The moral right of the author has been asserted.
Published 2023
Printed in the United States of America

32 31 30 29 28 27 26 25 24 23   1 2 3 4 5

ISBN-13: 978-0-226-82634-9 (cloth)
ISBN-13: 978-0-226-82635-6 (e-book)
DOI: https://doi.org/10.7208/chicago/9780226826356.001.0001

First published in Great Britain in 2023 as *Mountains of Fire: The Secret
Lives of Volcanoes* by Hodder & Stoughton
An Hachette UK company

Library of Congress Cataloging-in-Publication Data

Names: Oppenheimer, Clive, author.
Title: Mountains of fire : the menace, meaning, and magic of volcanoes /
    Clive Oppenheimer.
Description: Chicago : The University of Chicago Press, 2023. | First
    published in Great Britain in 2023 as *Mountains of Fire: The Secret
    Lives of Volcanoes* by Hodder & Stoughton. | Includes bibliographical
    references and index.
Identifiers: LCCN 2022054475 | ISBN 9780226826349 (cloth) | ISBN
    9780226826356 (ebook)
Subjects: LCSH: Oppenheimer, Clive—Travel. | Volcanoes—Popular works.
    | Volcanology—Popular works. | Scientific expeditions.
Classification: LCC QE521.2 .O66 2023 | DDC 551.21—dc23/eng20230302
LC record available at https://lccn.loc.gov/2022054475

♾ This paper meets the requirements of ANSI/NISO Z39.48-1992
(Permanence of Paper).

'How do we know a mountain? . . . by understanding its position on the globe . . . or by supplicating the gods that call the mountain home?'

– Ruth Rogaski, 'Knowing a Sentient Mountain'[1]

EASTERN HEMISPHERE

Eldgiá

Vesuvius

Stromboli

Tajogaite

Toussidé, Tiéroko & Trou au Natron

Nabro & Dubbi

Arctic Ocean

EUROPE

ASIA

AFRICA

Indian Ocean

Pacific Ocean

Pinatubo

Toba

Ili Boleng Tambora Merapi

AUSTRALIA

Southern Ocean

ANTARCTICA

WESTERN HEMISPHERE

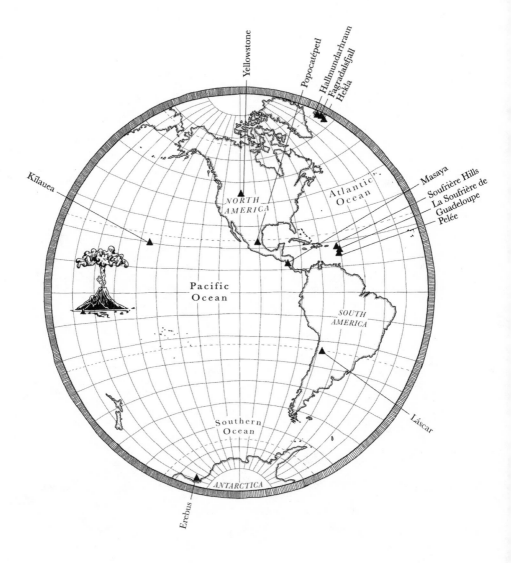

Yellowstone

Popocatépetl
Hallmundarhraun
Fagradalsfjall
Hekla

Kilauea

NORTH
AMERICA

Atlantic
Ocean

Masaya
Soufrière Hills
La Soufrière de
Guadeloupe
Pelée

Pacific
Ocean

SOUTH
AMERICA

Southern
Ocean

Láscar

ANTARCTICA

Erebus

Contents

Color illustrations follow page 230

Dreamland of the Living Earth

'The Fiery Gulph ejected great Stones'. Amerindians and
Spaniards meet at the foot of Popocatépetl, México, in 1519.[1]

*'Observations are the only truth ... Therefore, one good way,
probably the only way, to add to the scope of science is to make
reliable new observations of the unknown.'*
— Jack Oliver, 'Earthquake seismology in the
plate tectonics revolution'[2]

G rains of ash are dropping from the sky after a piercing detonation; they tinkle on my rucksack. Several large lava bombs blaze high above the crater, but fall back into the bowl-shaped depression, whose inner walls are striped with bands of cinders and rubble. At the bottom are two pits – one exhales a dusty smoke, but sometimes a much thicker brown cloud billows out silently to unfurl then dissipate across the larger, magma-filled vent, whose fumes are blue-tinged.

Trade winds sweep ash off the outer crater rim behind me like sand off a dune, showering the grey plain below, which stretches to a tract of lime- and emerald-green scrub and bush. The humid climate does a good job of confining the volcano's flagrant footprint. Ahead, beyond the crater, the South Pacific spans half the horizon, from silver under the sun to hazy blue.

Another deafening volley! I sense the heat on my face this time as lava dances up from the sloshing maw with a roar like the full thrust of jet engines – I feel it in my chest and through my feet. The crater fills with a wreath of sulphurous fumes; they spill out, making acid tears in my eyes. More ash prickles my skin. The experience is dynamic, elemental, mesmerising; it assails all the senses at once.

Now the fumes have thickened and I can no longer see where the bombs are flying. I turn off the spectrometer I've been using to measure gas emissions, and start heading back to my lodge in the forest. It's better to not sacrifice oneself for one's art.

★

Volcanoes get a bad press. They are most in the public eye when tourists have been assailed by lava projectiles, neighbourhoods buried beneath pyroclastic flows, populous shorelines ravaged by tsunamis, or planes grounded owing to the ash forecast. But volcanoes mean more than menace and calamity. Dramatic and traumatic as their outbursts can be, most volcanoes, most of the time, are tranquil mountains with diverse microclimates and habitats, and valuable mineral and geothermal resources. If we think of the places where humans have long lived in the shadows of volcanoes, the volcanoes were almost invariably there first. Like our parents, they've led whole lives before we get to know them. They are visual anchors in our landscapes and paint the sky with their plumage; they are supernatural realms; and they can turn the world's weather on its head. Even when their wild days are long past and their flames forever extinguished, their eroded landforms still enliven our skylines and invite outdoor adventure. Wherever we live on the planet, they are more a part of our lives than most people realise.

Volcanoes loom at a thrilling crossroads of nature, spirit, climate, geology, technology, society and culture. They play with time – stretching it over a geological epoch, yet able to shapeshift and change everything in the blink of an eye. As portals, they allow us to trace story and memory through deep time and back again.

As a volcanologist, I have dedicated my career to observing simmering craters, often at very close quarters, with a view to revealing their secrets. I've followed in the footsteps of pioneers like the American geologist, Thomas Jaggar, who established the Hawaiian Volcano Observatory in 1912. I love his description of geology being the 'science of the dreamland of the earth's interior',[3] and much of my work has involved recording phenomena at the mouths of volcanoes to help us understand their anatomy and physiology, to visualise their unseen lungs and alimentary tracts.[4] The truth is, I spend a lot of my time imagining the underworld, and comparing the quirks and frolics of different volcanoes. They never asked for an advocate but I am not alone in seeking to translate the language of these sonorous mountains for a wider audience.

Volcanoes are hard to ignore, especially if you live near one. We have probably admired and feared them ever since our species evolved in the shadows of Kilimanjaro and other fire mountains of eastern Africa, a few hundred thousand years ago. Given their sonic and visual spectacle, even between eruptions, it seems certain the ancestors would have sought to interpret their omens. But when did the more systematic study of volcanoes begin? Whose shoulders have I stood on in hope of seeing further? Historians of science might well differ on its origins, but I trace volcanology's first whispers back to the period when the term *volcano* was coined, and to the man whose careful observations would establish a template for centuries of colonial exploration (and exploitation): Gonzalo Fernández de Oviedo y Valdés.[5]

As a teenager in 1493, Oviedo must have felt the lure of the unknown when he joined a jubilant throng in Barcelona cheering the return of Christopher Columbus from the Caribbean. But before succumbing to the draw of far-off adventure, Oviedo settled in Italy, sensitising him to sulphurous volcanoes such as Solfatara, a steaming, bubbling crater near Naples, and Vulcano, one of the Eolian islands north of Sicily, famed since antiquity for its flamed frenzies. During this period, he also acquired a sensibility for figurative art, meeting Leonardo da Vinci, Mantegna and Michelangelo.[6] These experiences would serve him well as official chronicler – in both words and pictures – of the Spanish conquest of the New World.[7]

It was 1514 when Oviedo first sailed for the West Indies. His mission: to oversee the production of gold and silver. His compatriots, meanwhile, were expanding Spanish dominion on the mainland. Hernán Cortés landed in Veracruz (in modern-day México) in 1519, and within months was leading an assault on the Aztec capital, Tenochtitlan. Reaching the high pass between two towering volcanoes, Iztaccíhuatl and Popocatépetl, Cortés marvelled at their height and the 'flames' emitted from the latter.[8] The active crater of Popocatépetl would later furnish him with sulphur, a key ingredient of the gunpowder he needed to wage his campaign.[9]

By 1524, the Spaniards were colonising Guatemala and Nicaragua, where they encountered one volcano after another in various states of activity. This posed a problem – until now, there was no word 'volcano'; they were known instead as 'mouths of fire' or 'burning mountains'. Their superabundance in Mesoamerica cried out for the invention of a simple term. Many of the conquistadors were familiar with the Mediterranean sea. Like Oviedo, they knew of the island of Vulcano, itself named after the Roman god of fire, Vulcan. And so, the proper name seeded the generic *volcan*, which quickly permeated other languages.[10]

The most fabled volcano in all the region, and before long the subject of wild stories in European circles on both sides of the Atlantic, was Masaya in Nicaragua. The first colonial governor of the country, the brutal Pedrarias Davila, with whom Oviedo had sailed across the Atlantic, wrote of it in a letter to Emperor Charles V in 1525: 'there is a large mouth of fire which never ceases to burn and after dark . . . there is light as if it were day'.[11] Some said you could read a book by its brilliant glow at night, others that it hosted valuable reserves of sulphur. Holy men denounced it as the mouth of hell, among them the priest Francisco de Bobadilla who went so far as to exorcise the volcano, erecting a tall cross at the crater's edge.

But Oviedo wasn't someone to take the word of others unquestioningly, and so determined to explore Masaya for himself. He travelled there in July 1529, staying at a ranch near the village of Nindiri, a few miles from the peak. What happened next is recounted with brio and flourish in his fifty-volume *History of the New World*.[12]

Oviedo set off on horseback in the middle of the night, escorted by Nacatime, the Chorotegan chief of Nindiri, and an African servant. The trail across old lava flows, rough as 'blacksmith's slag', was tough. No shoe suited to it had yet been fabricated, in Oviedo's opinion.[13] When the incline became too steep for the horses, he put on wooden sandals and continued on foot.

The sun was rising as they reached the summit area: a plateau covered with multicoloured rocks and punctured by a chasm 'so wide a musket ball could not traverse it' and, from judging the descent of rocks hurled into it, around 130 fathoms (240 metres) deep. Oviedo spent the next hours applying all his senses and attention to observing, note-taking and sketching.

One of the puzzles he hoped to solve was the origin of Masaya's acclaimed radiance. At close range, he could see the light came not from flames but rather from fumes emerging from the funnel-shaped crater, which were lit up like smoke reflecting a bonfire's blaze by whatever burned deeper down. Leaning over the brink, he could make out a round floor far below, large enough to contain 'a hundred fencing cavaliers watched by a thousand spectators'.[14] Recessed into it was another pit, at the bottom of which seethed 'a fire liquid as water, the colour of brass'. It put him in mind of gold foundries he had inspected, except that it was covered with a black scum that repeatedly tore open to reveal matter 'red and brilliant as the light of heaven'. Occasionally, it surged into the air, plastering the walls of the pit. Since clocks were not yet commonplace, Oviedo marked time by reciting the *credo*. The expelled spatter glowed for six *credos* before extinguishing – around five minutes, by my reckoning.

Everything caught Oviedo's attention: the sounds like ocean waves dashing against rocks, the sulphurous greasy fetor of the fumes, the fierce radiation from the pit warming his skin. If his units of time, distance and area seem quaint, make no mistake: this is a luminous and faithful description of a lava lake. Furthermore, his keen powers of observation didn't end with the survey of physical manifestations. He asked Nacatime about a pile of broken pots, plates and bowls lying on the crater lip. The chief explained they'd contained food offerings for an ancient seeress with spiky hair, fiery eyes and razor-sharp teeth, who dwelt in the volcano. He would meet secretly at the crater's edge with other village heads to consult her on matters of war and peace, and to make oblations in anticipation of a good harvest. Oviedo learned that dishes of stew were not always sufficient tribute, though. During droughts or following earth-quakes and storms, women and children were hurled into the magma to appease the prophetess.[15] Around the world today, people still protect themselves from volcanoes with rituals and magic as much as with science and engineering, though thank-fully without resorting to human sacrifice.

Summing up his feelings, Oviedo wrote that no other volcano was as worthy of admiration or as remarkable, yet no Christian beholding it could be oblivious to the everlasting fire awaiting those ungrateful to God. This ambiguity cemented Masaya's reputation – as both marvel and hellhole – for centuries to come.[16]

Not all holy men feared the volcano, however. On reading Oviedo's testimony, a Spanish friar, Blás de Castillo, became convinced that what Oviedo had seen churning in the crater *was* molten gold. For months, he schemed with three accomplices, making furtive visits to the crater, before attempting to extract the precious liquor on 13 April 1538. 'Keep your silence,' the monk commanded the others. 'God doesn't want the gold to be discovered by the rich, but by the poor and humble.' Crucifix in hand, habits gathered at the waist, with an iron helmet for protection and flask of wine for courage, he climbed into a basket at the end of a rope, to be lowered into the crater by his henchmen. He reached the floor, but the 'gold' was still out of reach within a deeper pit. Three days later, the conspirators were back. This time, they set up pulleys to drop an iron bowl into the inferno, but it fused with the molten matter, and an exhausting tug of war ensued. More manpower was needed.

At this point, the friar confided in the provincial governor, who, a fortnight later, watched in disbelief as Blás and seven assistants were hoisted down into the crater. One can imagine a handful of Chorotegans watching the proceedings impassively. After much effort, they managed to haul out several samples before the chain attached to their lava scoop broke.

It didn't take long, of course, for the molten matter to congeal, and after all the effort, who would have dared suggest the residue looked an awful lot like the rocks underfoot? The samples were examined by the colonial mint in León.[17] The inevitable finding followed. An inflamed governor forbade Blás from any future prospecting, but the friar could not relinquish his dream so easily. He travelled to Spain, and successfully petitioned the emperor for the rights to reap Masaya's wealth. Sadly, he died soon after returning to Nicaragua. Some might think him avaricious, delusional, reckless, but he was a dreamer and fantasist of the highest order. He deserves a crater named in his honour.

Volcanoes, then, can be sites of encounter and performance, where history is made. Sometimes their eruptions are so dramatic they *make* history themselves – the Indonesian volcano, Krakatau, which exploded violently in 1883 claiming 36,000

lives, comes to mind.[18] And volcanoes are accomplished scribes
– they *write* their history, in the folios of pumice and ash from
which they are built. Pare back the layers of the archive and you
might find ancient soils, agricultural land, towns, footprints,
bodies and all traces of once-vibrant life whose chance exhuma-
tion connects us across millennia with fleeting seconds of peril.[19]

⋆

As Oviedo noticed centuries ago, volcanoes are a feast for the
senses. There are myriad attributes to the behaviour of a
volcano one could choose to record. But because it is the gases
in magma that influence whether a volcano erupts like uncorked
champagne or oozing syrup, I chose to dedicate a large portion
of my life to observing the fumes they emit. A Japanese chemist by the name of Sadao Matsuo once described volcanic gas
as 'a telegram from the earth's interior'. My work aims to
decode these vaporous messages to help me probe the dreamland, to seek answers to the mysteries of magma. I use a technique called spectroscopy to measure volcanic emissions, rely-
ing on the principle that every gas leaves its own unique
fingerprint on the light that passes through it.[20] The proportions of different gases – such as water vapour, carbon dioxide
and sulphur dioxide – can signal a volcano's mood, suggesting
what it might do next.

Making observations is where my abilities and proclivities
lie,[21] and much of my research has involved expeditions, a word
that originates from Latin, for 'to free the feet from fetters'. I am
a great believer in experience, experiments and experts (the
roots of these three words also derived from Latin, for 'to try' or
'to put to the test'). Models and projections are beguiling – they
readily spawn striking graphics purporting to tell us what the
future holds. But without grounding in real-world observations,

their value as tools for forecasting – whether it be of economic efficiency, a pathogen's progress, or a volcano's vivacity – is questionable. This isn't to say that making observations is the *only* way to do science – reason and experience should go hand in hand. I try then, while on fieldwork, to unencumber both feet and mind, and to do all possible to come home with good data. I've often found that putting in the groundwork is the best way to give serendipity a chance to play its hand and thereby learn things beyond my imagination.

Until recently, scientific knowledge of volcanoes was almost exclusively the domain of geologists, geophysicists and geochemists. Today, some of the most exciting research is being done at the edges of the discipline – where it meets anthropology, history, climatology, ecology and glaciology, to name a few. Volcanology is also an applied science – dedicated observers around the world keep a watchful eye on restless craters to protect their communities. During my career, I've watched the field broaden immensely in scope, gain confidence, lead the way in addressing wider challenges such as climatic change and disaster risk reduction, and make positive steps towards empowering the people it touches.[22]

But academic scholarship is not the only way to make sense of volcanic activity. Geologists now widely accept that knowledge accumulated through experience, wisdom, ancestral culture and spirituality matters.[23] Not only can it convey information on the nature and consequences of eruptions of the distant past[24] it also conditions how people respond to volcanic crises.[25] Imagine if unknown scientific experts suddenly show up in a neighbourhood, try to explain that their seismometers indicate an impending eruption, and ask everyone to abandon their homes and land. It would not be a recipe for successful risk mitigation. And this is why volcanologists increasingly interact with society, so that all sides are understood when the pot starts bubbling.

Volcanoes shaped my perspectives on causality, agency, risk and how knowledge is built. They taught me to listen: to indigenous tribal elders living close to restive craters; to experts in different branches of learning; to the lava breathing within our lively Earth; and to the messages, not meant for me, that make their way to the sky. Volcanoes changed me, and I believe strongly that they offer us all a different and unexpectedly humanizing way of seeing the world.

Land of God

Lazzaro Spallanzani observing Stromboli. Can you spot him?[1]

'All around us we could feel the power of an extraordinary world on a scale that was not our own, and absolutely indifferent to our existence.'

– Haroun Tazieff, *Craters of Fire*[2]

Her handbag stuffed with the money in one hand, and a suitcase in the other, Karin struggles to stay upright on the incline of volcanic gravel; each footfall dislodges a flurry of fine dust. She stumbles again as a gust steeped in acrid fumes from the crater engulfs her. She gasps for breath. Glancing back, she sees through burning eyes the village she has abandoned, clinging to the island's shoreline far below. Antonio must be searching for her by now. Tormented and exhausted, she unwittingly drops her possessions and clambers on – now, all she carries is her unborn son.

As night falls, she reaches the barren peak just as a thunderous detonation from the crater blasts fire and cinders into the sky. At the sight of such an infernal and meaningless abyss, Karin collapses, sobbing hysterically, her hot tears absorbed by black ash. 'Enough, enough!' she wails. 'God, if you exist, bring me some peace!'

God's response to Karin's petition at the close of Roberto Rossellini's film *Stromboli: Land of God*, is ambiguous, but for Ingrid Bergman, who played Karin, there was surely no peace. Her personal life at the time was as volatile as the cauldron of magma beneath the island, one of the seven volcanoes of the Aeolian archipelago that rise from the Tyrrhenian Sea between Sicily and the toe of Italy. Hollywood's biggest and most wholesome star of the 1940s was having an extramarital affair with the Italian director, filling gossip columns on both sides of the Atlantic, and shocking (and thrilling) a prurient public.[3] Perhaps the primal excrescences of the island aroused their adultery.

Bergman *was* carrying a child during shooting – is this why the denouement of *Stromboli* is so powerful?

But despite the film's sustained dramatic tension, the real-life scandal wrecked its release in 1949/50. Among those offended was Senator Edwin C. Johnson of Colorado, who condemned the 'stupid film about a pregnant woman and a volcano' on the floor of the U.S. Senate, describing Bergman as 'an apostle of degradation' on an 'immoral binge', and Rossellini as 'vile and unspeakable'. Other critics disparaged the director's use of unprofessional actors, including two sailors recruited for lead roles, and Stromboli's residents, cast as extras. It is true that the islanders were so baffled by the production that Rossellini tied strings to their toes so he could orchestrate their actions on camera. But the documentary style and visual aesthetic give the film an uncommon intensity and authenticity – in one scene, the villagers flee from a hail of lava bombs; in another, the sea erupts with tuna as the fishermen entrap and haul them on to their boats (much to Karin's disgust). Today, *Stromboli* is rightly regarded as a masterpiece of Italian cinema.[4]

After the Second World War, the island of Stromboli was an isolated and forbidding spot, one of the most remote parts of Italy. Mass emigration had emptied the villages, and a weekly mailboat from Naples was the only regular connection with the mainland. At the beginning of *Stromboli*, we learn that Karin has met Antonio in a camp for displaced persons. Marked by untold experiences of the war in Europe, she marries him when hopes of emigrating to Argentina are dashed. But she has not imagined the desolation and resentment that awaits her on Stromboli, where her feckless husband resumes his job as a fisherman. The confinement and alienation prove unbearable for Karin, but her ultimate confrontation is with the utter callousness of sulphurous nature: the volcano.

★

Rossellini's epic was not the first volcano movie. Fifty years earlier, the British filmmaker and pioneer of animation, Walter Booth, had made a short adaptation of Edward Bulwer Lytton's hugely popular novel, the *Last Days of Pompeii*.[5] We might even see the roots of volcano cinema in the *Vesuvian Apparatus* of Sir William Hamilton, Britain's envoy at the court of Naples from 1764 to 1800, and a man who did more than most to excite public interest in Pompeii and Vesuvius. His contraption consisted of a wooden cabinet with a transparent coloured painting of the volcano at the front. Inside, a clockwork mechanism presented flickering light to the lava flows swirling down the volcano's flanks, while striking a drum to simulate the accompanying booms.[6]

But more than a designer of drawing room entertainments, Hamilton was the first serious volcano watcher in Western history. The timing of his posting to Naples could not have been more propitious to turn a savant to volcanic affairs. Vesuvius was fired up for a prolonged bout of pyroclastic vigour, while excavations at Pompeii and Herculaneum were yielding spectacular relics that had been entombed in pumice for seventeen centuries. Hamilton compulsively acquired the spoils of the archaeological digs, but also made more than sixty ascents to the crater of Vesuvius over a period of thirty years, collecting rocks and minerals, and documenting eruptions and the transformations they effected on the landscape. He travelled with an entourage of assistants and artists who sketched and painted the eruptive phenomena. On one visit, in 1767, he and a guide were caught in an eruption. In near total darkness beneath thick ash clouds, they hotfooted non-stop for three miles to outrun a torrent of lava coming at them. In an understated report of the episode, Hamilton noted the ejecta to have been of 'such a size as to cause a disagreeable sensation upon the part where they fell', and that it had been 'prudent' to retreat. So strong was the sustained seismic shaking when he reached his villa that he gathered his household and retreated to Naples, stopping at the royal court in Portici to urge King Ferdinand to flee, too.

Lava flows threatened several towns, and the sustained roaring from the mountain incited widespread terror. People opened umbrellas to shield their faces from falling ash. Even Hamilton, mindful of the terrible eruption of 1631, which had claimed 4,000 lives in Portici, feared 'some dire calamity'. It was time for San Gennaro, the preeminent patron saint of Naples, to intercede. His effigy was processed towards the volcano, subduing its clamour. At least, this is how the public viewed it, though Hamilton was unconvinced – he had observed many alternations in the crater's sonority before.

Hamilton was a renowned socialite, and anyone of distinction or erudition would call on him if in Naples. And many came, attracted by the region's classical heritage and vivacious volcano, a titillating highlight of the Grand Tour, its ascent a rite of passage for young aristocrats. Goethe was among those aroused by sublime encounters with gushing lava. He also got on famously with Hamilton, whose universal taste he admired. With this cultural vortex centred around Naples, Hamilton was swept into contemporary scientific debates and successfully built connections to the learned societies.[7]

The perfect marriage of Hamilton's dual passions for antiquities and volcanism is represented in his first-hand report *Campi Phlegræi: Observations on the Volcanos of the Two Sicilies*, printed in instalments from 1775. Hamilton nails his empiricist colours to the mast in its opening discourse, writing: 'Accurate and faithfull observations on the operations of nature, related with simplicity and truth, are not to be met with often.' Going further (and echoing Oviedo's unwillingness to accept hearsay), he laments that 'those who have wrote most, on the subject of Natural History, have seldom been themselves the observers, and have too readily taken for granted sistems, which other ingenious and learned men, have perhaps formed in their closets'.[8]

Fidelity, then, mattered to Hamilton – at least in scientific practice (he famously tolerated his second wife's notorious affair with Horatio Nelson). In *Campi Phlegræi*, he presented the most authentic depictions of volcanic activity, geological strata and rock specimens possible. The fifty-four plates in *Campi Phlegræi*, 'colour'd after Nature', were made by the artist Pietro Fabris under his vigilant supervision. This was a radical departure: scientific treatises of the period were usually embellished with stylised monochrome engravings, characterised by fanciful juxtapositions, impossible gradients and whimsical renderings of the subterranean.[9] So sumptuous and costly were Hamilton's illustrations that, despite a favourable public reception, publication of *Campi Phlegræi* actually added to his enduring financial precarity, born of lavish entertaining and profligate collecting.

Campi Phlegræi set a benchmark for science communication. I admire it greatly for challenging the (still commonplace) narrative that volcanoes connote doom – Hamilton hoped his 'exact representations of so many beautifull scenes' would help people see volcanism in 'a Creative rather than a Destructive light'. And his work influenced art as well as science, notably a later generation of landscape painters that included Caspar David Friedrich.[10]

And yet, despite his passion for all things igneous and his true-to-nature approach to scientific visualisation, Hamilton's volcanological legacy is eclipsed, for me, by his contemporary, Lazzaro Spallanzani, a priest and professor of natural history at the University of Pavia in northern Italy. In fact, Spallanzani is much better known as the founder of experimental biology; it was only late in life that he turned to the volcano.[11] It is his experimental as well as observational approach – and his eye for research design – that marks him out. He didn't just collect rocks for the university's cabinet of natural history: he pulverised, heated, chemically treated and analysed his samples to simulate processes he had witnessed beside the crater. All the while he posed questions and probed assumptions in his pursuit of nature's secrets. His methods feel extraordinarily modern, and are all the more remarkable considering that geology barely registered as a discipline in the late 18th century.

Well acquainted with *Campi Phlegræi*, Spallanzani gave 'the most attentive consideration' to Hamilton's work, but found it not uniformly 'consonant with fact'. The British diplomat lacked mineralogical knowledge, he wrote, and was ready to accept accounts of others 'more marvellous than true'. Hamilton's opinions on Stromboli were particularly flawed, he added, being based on observations made at sea with only the aid of a telescope. While other proto-geologists of the period hardly set foot in remote outposts like Stromboli, as an energetic traveller who knew how to plan a field mission, Spallanzani would make his inspection at disconcertingly close range. This valour was noted by the Swiss scientist and eulogist Jean Senebier, who wrote: 'He dissected, as it were, the uninhabited volcanoes, with the exactness of a naturalist anatomizing a butterfly, and the intrepidity of a warrior defying the most imminent dangers'.[12]

Spallanzani arrived on the island in October of 1788 and set about climbing the volcano at once, assisted by local guides. But his progress was thwarted when he reached 'the extreme edge of the smoke', which resembled ordinary cloud, except that it was, as he subtly recorded, 'inconvenient to respiration'. He retreated down the slopes to the coastal village, where he lodged with an elderly priest. Just a stone's throw from the crater, the house was rattled by explosions, which Spallanzani correctly attributed to atmospheric shockwaves rather than seismic shaking. In the morning, a sprinkle of black ash settled upon the village.

Two days later, under more clement skies, the tall and well-built Spallanzani set out again, dressed in a fine coat, waistcoat and breeches, and sporting a wide-brimmed, flat-crowned hat. The going was good on this second attempt on the volcano – until he reached a vast scree of cinders strewn with occasional larger boulders of lava. This engulfed him up to his knees with each stride. Just a few months shy of his sixtieth birthday, he must have been a fit and adventurous man for his age to endure and enjoy this taxing ascent. At 900 metres above sea level, he came to a precipice known as the Pizzo sopra la Fossa, the 'peak above the pit'. Stromboli's active crater lay at the far end of a wide, bowl-like depression stretching before him.

The gravelly surface at his feet was strewn with big clots of unweathered lava. Spallanzani correctly surmised they could not have rolled there, and that their fresh appearance suggested recent arrival. They had to have been hurled through the air. Undaunted by the prospect of further ballistic projectiles, Spallanzani marched towards their point of origin for a better view. Luckily, he found a small recess below the Pizzo, protected by an overhang. In an illustration from his published research, he is secreted in this nook, at one with the grain of the volcano, and staring into the fire pit. What he saw there reminded him of a vat of molten brass. The lava would rise, its surface blistering until large bubbles ruptured it violently, spurting red-hot rock into the air. Now the magma would subside until the next upsurge of fiery spume. For Spallanzani, nothing less than 'Truth and Nature' were exposed before his wondering senses.

While Hamilton had noted Stromboli's intermittent and monotonous activity, Spallanzani could see that, in fact, the lava seethed ceaselessly; louder explosions expelled more rock. But just when he thought he recognised a pattern, the lava level plummeted, and the vent silenced. Then, jets of incandescent fume burst from apertures along a fissure that was unnervingly nearby. Spallanzani feared these vapours might have 'mischievous effects', and was about to leg it when his guides reassured him there was no threat. They said the 'air' animating the lava had temporarily found an alternative path to the surface, and that normal activity would resume shortly. Spallanzani was duly impressed when the prediction was borne out.

Displaying a respect for local lore not widely shared by naturalists of his day (most of whom were scathing of anything that smacked of 'peasant superstition'), Spallanzani was eager to discover more of the islanders' knowledge of their volcano. He learned that the crater's fuming foretold the weather – when subdued, more northerly winds would prevail, whereas increased vigour of the plume heralded southerlies. The great Roman philosopher, Pliny the Elder, had heard this, too, writing in his monumental *Natural History* from 79 CE: 'From the smoke of this volcano it is said that some of the inhabitants are able to predict three days beforehand what winds are about to blow.' Even further back in time, the Greeks regarded Stromboli as the home of Aeolus, the god and keeper of winds. In the *Iliad*, Odysseus was entertained at Aeolus' palace on a 'floating island' with sheer cliffs and a 'wall of unbreakable bronze'.[13]

★

It was only natural to begin my research career in the cradle of volcanology. Two hundred years to the month since Spallanzani's fieldwork on Stromboli, I followed in his footsteps, though not nearly so stylishly costumed. I know exactly how his legs ached as he slogged up the side of the volcano, and how he must have admired the spectacle of its internal commotions.

Stromboli's main settlement clings to the north shore of the island. It was a fine day as I set off, but, being out of season, there were very few tourists about. Many of the whitewashed cellular homes were shuttered, front gates padlocked, bougain-villea and hibiscus flowering untamed over garden walls. The trail led out of the village, then meandered across abandoned terracing, a relic of the viticulture for which the island was once famed before the destructive pest, phylloxera, arrived in the late nineteenth century. The bugs ravaged the grape vines, provok-ing a great exodus from the island. Another wave of emigration followed in the 1930s, so that within two generations the popu-lation declined from several thousand to a few hundred.[14] This abandonment is evocatively portrayed in Rossellini's film, and represents the cultural backdrop to Karin's profound isolation on Stromboli.

Nearly halfway up, the trail reaches the margin of an abrupt declivity on the northwest flank of the volcano – the Sciara del Fuoco, which roughly translates as 'the scar of fire'. This wound, formed by an ancient landslide of part of the cone, plunges to the sea. It is the chute down which the vents spill most of their lava. The submarine part of the volcano extends twice as far beneath the waves as it projects above; most visitors see Stromboli as a modest volcano in scale, but strip away the Tyrrhenian Sea and it would not look out of place with its loom-ing cousins in the Pacific Ring of Fire.

Nearing the peak, I witnessed my first eruption. It was brief but thunderous, and accompanied by a convoluting puff of brown ash that had punched through a garland of white steam clouds gathered about the crater. The mottled plume rose skyward until the wind sheared it off towards the sea. Loose cinders underfoot made the rest of my ascent more of a slog, like climbing sand dunes. But twenty minutes later, I had reached the Pizzo sopra la Fossa. I could now make out a row of four cones, shaped like upturned dishes and built from accumulations of dark volcanic slag. Each was truncated by a fuming crater, the closest of which greeted me with an arresting whoosh. I had arrived in a pristine world that I could not have properly imagined. Nothing felt stable; all was sterile, temporary, shifting, piling up, subsiding.

What had brought me into this world above the clouds? I was in my twenties and had started the first year of my PhD with a research trip to Southern Italy. Armed with little equipment apart from a notebook and a thermometer capable of readings over 1,000°C (1,800°F), my quest was to investigate whether it was possible to take the temperature of a volcano from space. The idea of monitoring volcanoes from Earth's orbit was not new. As early as 1966, one of the first weather satellites, Nimbus II, had registered the heat radiated from lava flows on the newborn island of Surtsey, south of Iceland.[15] By the mid-1980s, with more advanced infrared imagers spaceborne, it looked possible not only to detect eruptions but also to quantify areas and temperatures of lava or hot gas vents. Underpinning this approach was a foundational equation relating the spectrum of heat radiation from a surface to its temperature, derived by the physicist Max Planck in 1900. As well as expressing 'red-hot' in physical units, the law explains why a red-hot surface glows red-orange, orange and then yellow to white as it heats up. But several assumptions were needed to apply Planck's formula to volcanic features, and my task was to test them in the hope it would pave the way to an orbiting volcano observatory. This required 'ground-truth' – lava temperatures measured at close range to compare with measurements made with the satellite data. With hindsight, I got unduly close.

At the top of Stromboli my main tool was an infrared radiometer, a 'telephoto' version of an ear thermometer. It resembled a video camera and, unlike its clinical counterpart, did not require direct insertion into the orifice – I could record temperatures at a distance. I particularly wanted to investigate the uniformity of lava surface temperatures and their variability over time. But on reaching the Pizzo sopra la Fossa, I faced the same quandary as Spallanzani two centuries before – the lava vents within the craters were hidden from view. I needed a direct line of sight. And for that, I had to get nearer.

Nothing like Spallanzani's niche caught my eye, so I loped down the southern rim of the wide depression and then up to the lip of the nearest crater. I leaned over the chasm. The inner walls exposed layers of lava and scree tinged pink from oxidation, or stained yellow where sulphur had condensed around innumerable perforations in the rock. At the bottom, surrounded by drifts of rubble, two magma-filled pits sputtered and wheezed noisily, spitting bright-orange molten rock from their mouths every few seconds. But at longer intervals of ten minutes or so, larger explosions would blast projectiles well above my level. Some of these detonations sounded like pistol shots or thunder-cracks; others roared for ten or twenty seconds. Each would be followed by a succession of dull thuds as a fusillade of solidifying lava fell to earth, followed by a clatter like hailstones on a tin roof as smaller fragments rained out – an igneous rendition of *musique concrète*.

Ignoring the tumult, I fixed the infrared thermometer on a tripod and trained its optics at the most expressive vent, which was also the closest. Unfortunately, the device had no inbuilt recording capability, so I had to read the temperatures in the telescopic viewfinder, and scribble them down in a notebook as fast as possible, while simultaneously keeping an eye on my watch to time the observations. With a little practice, I managed a measurement every second or two.

Through the viewfinder, I could observe the live action of the blusterous vent. The explosions began with a tumescence of the lava, much as Spallanzani had described. The magma skin would then rupture, unleashing a jet of blue flame. An instant later, a blast of orangey-red glowing spatter and bluish fumes shot above the mouth of the crater. Soon, I would experience a light dusting of dark glassy shards and crystal flakes. They collected in the crease of my notebook, making it easy to decant small samples of the ash – just forged in the fiery aperture – into plastic bags. The explosions reverberated in my chest, seismic tremors shivered my feet, and a metallic whiff like a struck match smarted my eyes. I was seized by the realisation that volatile molecules just unfettered from the inner Earth, and tasting like sour milk at the back of my throat, were now in my lungs, in my bloodstream.

In fact, these kinds of exhalations are exactly what established the Earth's primary atmosphere aeons ago. We are made of the very same stuff, and some think that life began in the primordial chemical soup of volcanic vents. Through the actions of tectonic plates and volcanoes, our fleshly inventories of carbon, oxygen, hydrogen, nitrogen, sulphur and many other elements are making just a brief sojourn in an eternal cycle between the deep interior and the surface of the planet. I was discovering, as I am certain Spallanzani felt, that fieldwork on an active volcano is a profoundly embodying experience.

Another booming volley! But this time lava spray reached not only above, but over the crater's edge. One piece landed not far away and I dashed over to inspect the missile. It was light grey with a frothy texture. The vernacular for this substance – *scoriae* – comes from the Greek word for dung. Any projectiles brick-size and up are known as 'lava bombs'. If they splat when they hit the ground, they are 'cowpat lava'. As my fingers closed on the find, I howled in pain: in my spontaneous curiosity, I'd overlooked the heat capacity of rock. My thermometer confirmed the surface temperature was hotter than a pizza oven. I look upon this period as the steep part of my learning curve as a volcanologist.

Back at my station at the lip of the crater, the activity was ramping up. Large explosions now recurred every few minutes. The thermometer readings plummeted as the crater filled with fumes. But I could still hear the polyphony of hisses, roars, rumbles and cannonades. When not making temperature measurements (with one hand still throbbing from the burn), I recorded the time and characteristics of each explosion. For one of these entries in my field notebook, the ink jerks up halfway through a word. I retrospectively labelled this jolt 'blind terror'. Half a minute later, an even more violent spasm daubed the crater walls with fluid basalt, and the subsequent salvo hurled molten cowpats at a lower angle, dropping one just a few paces away – it was still glowing. My notebook reveals growing risk acumen: 'Working here extremely hazardous,' I remarked.

The crater cleared, revealing again the smouldering vents. Like Spallanzani, I tried to imagine what lay beneath – did these pits converge on a common magma source? It puzzled me that two adjacent openings could display such different behaviour: one blazed steadily but seldom shed sparks; the other was erratic, boisterous and violent. As each second passed after an explosion, I tensed up in expectation of the next. The anxiety must have got to me, for I suddenly needed to eject my own *scoriae*: not the most convenient time or place. A double volley (from the volcano) propelled bombs well above me again, and this time I had to watch where the buzzing projectiles were headed in case it might be necessary to hop out of harm's way.

Then Aeolus switched tactics: it began raining. Cold, wet precipitation somehow caused me greater alarm than a threatening hail of lava bombs. I retreated up the ridge leading to the Pizzo sopra la Fossa to find cover. Several crude shelters dotted the rise. They were no more than knee-high, circular walls of lava rocks, but they afforded protection from the wind.[16] I found some plastic sheeting in one – the vestige of a tourist's adventures, no doubt – and stretched it across the top of the wall, then crawled under. For the first time, it struck me that I was completely alone on the volcano.

Even after the rain eased, there was no point in returning to the craters, because thick, white steam clouds, provoked by the wetting of hot rocks and magma, now filled their mouths. It would be impossible to see in, let alone take temperature readings. At dusk, these clouds reflected the lava light, becoming luminous like bonfire smoke. Explosions intensified the glow moments before incandescent lava bombs flared out, tracing ballistic arcs across clear sky – a mesmerising sight. In the distance across the sea, points of light in the darkness mapped the contours of settlements on nearby Salina island and the shoreline of Reggio Calabria on mainland Italy. It was hard to tear myself away from the pyrotechnics – 'Just one more,' I kept saying to myself – but eventually, I turned in for the night.

I slept fitfully in my crude tent, discomfited by the cold and disturbed by the incessant flapping of the sheeting in the wind. Acidic gases drifting from the craters had fumigated the ridge intermittently, and it had turned out I had a companion: my makeshift shelter was also the refuge of a rodent, which scuttled across me and sniffed my neck. At first light, I was about to tuck into *panini* and ricotta for breakfast, when I spotted neat puncture holes in the crust, through which all the soft interior of the rolls had been extracted. At that moment, a small mouse poked its snout from a crevice in the rocks.

My stomach bubbling from hunger, I returned to the crater to collect more data. But the fumes were too thick, and explosions sprinkled me with pea-sized pieces of lava froth. So I headed off to investigate the adjacent crater, which seemed to have been tranquil. Five minutes later, an ear-splitting explosion pelted my measurement site with ejecta. Reaching the rim of the unexplored crater, I found it, too, strewn with lava bombs.

Pondering the implications of a red-hot cowpat landing on my head, I felt a surge of unease and retraced my steps. I hadn't gone far when a massive discharge from this crater assailed the spot where I had been standing. Before there was time to reflect, an explosion from another crater propelled rocks high over-head, which buzzed as they picked up speed on descent. Feeling increasingly like a target in a shooting gallery, I called it a day and hiked back to the village.

*

It's an understatement to say now that this fieldwork was impru-dent. Given that lava fragments leave the vent at bullet-like velocities, I would have had no time to react had one been coming at me. It wouldn't have just been the end of my PhD studies.

I lived and learned. But even my antics pale in comparison with those of one of the most renowned and controversial volcanologists of the twentieth century, Haroun Tazieff, who explored the volcano with surprising impunity in 1949. Tazieff, of Polish descent, was very well known in France, where he settled, thanks to his thrilling films of expeditions to remote and intriguing volcanoes of the world. He was to volcanoes what Jacques Cousteau was to the oceans.[17]

Tazieff did not just go up to the active craters on Stromboli, he scrambled round the back of them, teetering along a narrow hinge between the chasm and the swooping incline of the Sciara del Fuoco. Dodging lava bombs, he came abruptly to a small opening in a subterranean passage filled with flowing magma, a feature known as a 'skylight'. Shielding his face from the heat, he aimed a handheld pyrometer (a forerunner of the device I used). It read 1,100°C (2,000°F). About halfway down the Sciara, he spied the lava re-emerging as a river. But this wasn't the time to explore it.

Down in the village, there was quite a commotion owing to the presence of Rossellini's production team. The director and Bergman were at sea filming the tuna-catch scene, and Tazieff befriended the film's location manager, General Ludovico Muratori. At forty-eight years old, Muratori was a veteran of Mussolini's war in Ethiopia, where he had commanded guerrilla units fighting the British and Haile Selassie's forces. Hearing of Tazieff's escapade on the volcano, Muratori asked where he might find a 'sensational, but safe' spot at the summit to shoot the film's climax, when Karin attempts to traverse the island by way of its magma pits. Tazieff suggested locations where the danger was 'minimal' – at least between explosions, he cautioned. For himself, he had a maximally hazardous adventure in mind: to reach the lava flow he had spied the previous day. And he wanted to observe it by night for greatest effect.

It was already dark by the time he was back at the brink of the Sciara del Fuoco, but this made it easier to chart a course to the lava channel, now a bright beacon. He slid down the shifting scree, anxious at many points of triggering a rocky avalanche that would sweep him away. He stumbled over boulders, one of which turned out to be a dull-red glowing block of lava that had spalled off a solidifying flow above. Fresh chunks of it broke free and bounded furiously downslope, casting off sparks with each bump. At another crag of lava, Tazieff tripped, smashing the lamp he had brought for the return ascent, which by now he was dreading. He was parched, and it took several more hours to reach the lava torrent. But he was spellbound by the inexorable blazing pageant, which reminded him somehow of columns of marching ants, and admired it for a while before getting out his instruments and cameras. The climb back, without a light, was every bit as horrendous as he'd feared, and he was desperately thirsty. There were several further narrow escapes: a bellowing rockslide; stifling clouds of dust and fumes; a broadside of bombs just paces away. But at last, he made it to the tent he had pitched near the top, where he had cached water.

Sunlight woke him in the morning, and he was enjoying a mug of tea when he spotted mules emerging from the sulphurous miasma encircling the peak. They were being led, but for one, whose rider lay at an awkward angle. Tazieff ran to the group and found the passenger was Muratori, who had come up to establish camps for the filming. His face was livid, his voice rattled, and he slumped against Tazieff, gasping, 'Fumes–air!' His condition deteriorated alarmingly, and Tazieff and the muleteers lifted him and started down to the village as fast as they dared on the dusty trail. But they did not get far before the general died in Tazieff's arms.

Was Rossellini's film worth a life? The general's death renders the scene where Karin is suffocating in the fumes very close to the bone. Bergman herself paid for the funeral and made a considerable donation to the bereaved family.

Were Tazieff's data important enough to justify such exceptional risks? My own observations at the crater's edge were to little purpose: futile, even, in retrospect. The thermometer readings were erratic – they rose and fell by the second, according to the ebullition in the magma vent and quantity of fumes passing across the instrument's view. I realised a satellite image snapshot of such a scene could do no more than confirm a volcano's activity – it would not yield a meaningful temperature measurement. This meagre finding merited a paragraph in a paper that only a handful of people have cited in the three decades since it was published.[18] I doubt that anyone who read it imagined the caper of collecting the data, unless they stopped to think on

reading that I had made the measurements at 'viewing distances
of approximately 50 metres'. Tourists are no longer permitted
anywhere near the craters. Even though my excursion happened
before the era of risk assessment, most people would have taken
one look at the bombardment and had the sense to keep well
clear of the evident peril. For me, it took a while for fear to
trump fervour.

*

The banality of my findings aside, the volcano of Stromboli is
undeniably among those that have propelled understanding of
the fundamental drivers of eruptions. It ranks in the top three
most-studied volcanoes worldwide, along with its compatriots,
Etna and Vesuvius. In his research, Spallanzani posed two questions that remain as relevant to volcanology today as they were
then: why is Stromboli's activity so persistent, as suggested by
records stretching back 2,500 years? And what is the explosive
mechanism?

On the first point, Spallanzani argued that whatever nourished Stromboli's perpetual conflagration was the same
substance animating other volcanoes known only for sporadic
eruptions. Some naturalists of the period regarded petroleum as
the energy source. They argued that this explained the dark
plumes that roared into the atmosphere during eruptions.
Spallanzani, however, discounted this hypothesis, believing
instead that an inexhaustible supply of burning sulphur, a well-
known active ingredient of gunpowder, sustained Stromboli's
incendiary behaviour. This would explain the stench of exhaled
fumes and the abundant yellow sulphur encrusting gas vents. As
to the nature of lava, while many of his contemporaries considered it to be a mixture of sulphur and bitumen, Spallanzani
insisted – correctly, as now seems obvious – that it was the

molten form of the rock found everywhere at the surface of the volcano.[19]

Now, we understand the generation of magma that feeds Stromboli to be a by-product of the sustained collision between the African and Eurasian tectonic plates. This induces localised melting of the rocks of the Earth's mantle, deep below the surface, producing magma that percolates upwards. It accumulates at a depth of around six miles, forming a chamber. This feeds a shallower 'reservoir' a little below the level of the seabed, and it is this that supplies the crater vents.[20]

Spallanzani tackled the question of Stromboli's activity systematically through innovative experiments on his return to Pavia. He was convinced that an 'elastic fluid' (in other words, a gas) must animate the magma column. After all, he had directly witnessed bubbles rupturing in the magma pond, and the hissing of pressurised gases from rents in the crater. To test his idea, he crushed each rock specimen to a powder, decanting some into a clay crucible. This was attached by glass tubing to a flask of mercury that would trap any gases expelled when the samples were heated. Looking through the end of the glass tube, Spallanzani observed the miniature magma cauldrons he had created, transporting him back to Stromboli.

Most of the captured gas proved to be air from inside the vessel, but Spallanzani recovered white fumes given off during some experiments. Tasting the condensed droplets, he identified hydrochloric acid, but was sure other gases must also be 'imprisoned' in the molten rock beneath Stromboli until they escaped thanks to their buoyancy. Reaching the confines of the vent, he argued, they formed the large bubbles he'd seen burst so violently, flinging out lava shrapnel.[21]

Spallanzani wondered about collecting the gas emissions on the volcano, but ruled it out because of 'the manifest danger there would be of falling a victim to so impudent a curiosity'. (I must add 'impudent' to 'imprudent' to describe my first field mission to Stromboli.) However, he reasoned, they were probably similar to emanations from docile volcanic vents at Solfatara near Naples, which were then known to contain 'carbonic' and 'sulphureous' acids and 'oxygenous gas'. Somehow foreseeing the fruits of twentieth-century technologies, he concluded: 'Of the existential nature of these substances, perhaps, we may not always remain ignorant.'

We now know that nearly two-thirds of the volume of the bubbles that burst in the vent consist of water vapour and a third is carbon dioxide.[22] Minor quantities of sulphur dioxide and hydrogen chloride (which Spallanzani had detected in his experiments) impart the plume's acidity. A bubble the size of a ping-pong ball deep in the volcanic plumbing becomes yoga-ball-sized near the surface, simply due to decompression. At the end of their journey, approaching the atmosphere, the expanding bubbles want to speed up, but their passage is resisted by an abundance of crystals in the magma.[23] Bubbles also catch up with each other, and coalesce to make larger bubbles that squeeze against the walls of the magma conduit. Reaching the vent, they distend the lava surface until it ruptures violently. The sudden acceleration of escaping gas eviscerates the molten rock, slinging fragments up into the air. A typical explosion discharges a tonne of gas and a few tonnes of molten rock – in total, as much as the weight of an average car. Videos of the explosions recorded at 500 frames per second and played back in slow motion have further revealed that each explosion is a composite of discrete pulses – there is not one bubble, but a whole procession.[24] When we change the scale of observation, so often we find something that we hadn't noticed before. Without bubbles, magma would be boring.

Spallanzani's research marks him out as a trailblazer in experimental petrology, a field that continues to illuminate the internal action of the Earth where probes cannot penetrate.[25] Just as Spallanzani tested his concepts for the underlying drivers of volcanic action through experiment, so today the metabolism of magmas can be simulated in the laboratory. Now it is possible to conduct such experiments both at high temperature, as did Spallanzani, and at the crushing pressures corresponding to the depths of real magma bodies. In parallel, the technologies for analysing the chemical and physical properties of samples have evolved tremendously.

Spallanzani's conclusions were influential for years to come and remain remarkably aligned with thinking today. Highlighting his importance, he is connected with the first recorded usage of the term 'volcanology' in English, in the June 1800 issue of *The Edinburgh Magazine*:

> We find in the Voyages of Spallanzani a new volcanology; he therein teaches the way to measure the intensity of the fire of volcanoes, to glance at the causes, to touch almost, in the analysis which he makes of the lava, that particular gas which resembling a powerful lever, tears from the bowels of the earth . . . those torrents of stone in fusion . . .[26]

His contribution was trumpeted again in the first English-language synthesis of volcano knowledge, published in 1825: *Considerations on Volcanos*, the work of a precocious polymath and politician, George Poulett Scrope. He'd visited Stromboli a few years earlier and wrote that Spallanzani was the first to establish 'the nature of volcanic agency in its true light'.[27]

On mainland Italy, Vesuvius continued to steal the show. The visiting scientists included Humphry Davy, discoverer of sodium and potassium, and inventor of the miner's safety lamp, who mistakenly attributed volcanic fury to the exposure of reactive metals to water and air. He attended several eruptions of the

Neapolitan volcano in 1819 and 1820 and hurled assorted chemical compounds on flowing lava but the experiments failed to validate his theory.[28] Meanwhile, local scholars began to lament the piecemeal attention paid to the volcano's behaviour – the activity was so changeable, it would be easy to miss something important simply because no one was present to observe it. Among those who saw the opportunities of continuous vigilance were the mineralogists Teodoro Monticelli and Nicola Covelli, who wrote in 1823: 'If educated men watched from a meteorological-volcano observatory to record all the events of Vesuvius, and observe their effects [. . .] a veil would be lifted on the physics of volcanoes'.[29] Without such a facility, they argued, making sense of Vesuvius would be like reading a novel from a selection of randomly detached pages.

Their dream was realised with the founding of the Royal Vesuvius Observatory in 1841 – the world's first dedicated volcano institute. It was built on a prominent cone on the slopes of Vesuvius with a fine view of the summit crater. But this was a time of political upheavals and its first director was soon ousted in a purge of intellectuals. It was several years until a successor, Luigi Palmieri, was appointed but he would hold the post for forty years. He designed and operated the world's first electromagnetic seismograph and, just as importantly, he ran an open-door policy, enabling collaboration with other field and laboratory scientists, and he regularised data archiving and reporting. With such organised and institutionalised inquiry, 'volcanology' had come of age.

The rising star of the discipline as Palmieri grew old was another Italian priest and professor, Giuseppe Mercalli. Like Hamilton and Spallanzani, Mercalli believed the key to understanding the natural world was to observe it.[30] It was he who coined the term 'strombolian' to describe the typical activity of Stromboli: moderate explosions of expanding gases that blast out projectiles in a persistent cycle. Strombolian activity is one of the most ubiquitous varieties of volcanic eruption – it is likely even to be responsible for deposits and landforms seen on Venus and Mars, albeit with changes of scale reflecting different surface conditions.[31] Mercalli also classified more explosive varieties of eruption: 'vulcanian', 'plinian' and 'pelean'. After many years yearning for the top job in volcanology, he was in his sixties when he became director of the Vesuvius observatory in 1911. But his tenure was tragically cut short when he was consumed by a blaze at his home. Some thought he might have knocked over an oil lamp and been unable to contain the resulting conflagration, but others suspected he was the victim of a botched burglary, the fire lit to conceal evidence of murder.[32]

Today, Stromboli still feels like the periphery of Europe, at least outside of the tourist season. The resident population remains just a few hundred. For farmers and fishermen, the volcano is a companion. But the relationship goes deeper. On one visit, while disembarking from the boat that had brought me from the neighbouring island of Lipari, I encountered one of Stromboli's resident priests. He was carrying a plastic supermarket bag, and I could not help noticing a glass case projecting from the top of it. It housed a silver gauntlet, a priceless reliquary allegedly containing remains of the Apostle Bartholomew, who is said to have been miraculously washed ashore on Lipari.[33] The priest confided that his parishioners' beliefs melded Catholic faith with traditional culture: 'They believe this is the land of God but also, that the land *is* God. Even now, when the volcano erupts more violently, the people shudder. It reminds them that another divinity pervades the island.'

I have encountered such accommodations of seemingly disparate creeds all over the world, wherever there are volcanoes. On Stromboli, the priest's observation was borne out when I met one of the island's oldest inhabitants at his home in the village. He had been in his teens when Rossellini and Bergman were filming there. On a dresser in his dining room sat a small shrine adorned with Christmas lights and housing keepsakes interconnecting the sea, Church, family and volcano: a framed tinted photograph of his mother; a model sail ship; a diorama of the village made with shells and lava stones; a souvenir snow globe from Rome; a nugget of sulphur beside a picture of the Madonna; and a plastic amaryllis. He asked if I would like to hear a poem he had written.

His weathered eyes twinkled as he recited (in regional dialect), and I found myself transported to Pizzo sopra la Fossa at dusk:

Dear friends, come, come!
I bring joy:
A gilded sunset, a million-coloured rainbow between sky and sea.
By night you hear me tremble.
But I do it to make you embrace.
Come, kiss me,
You won't forget it; my kiss is made of fire.
See the sparks burst from my mouth.
I hear you cry 'che bello, che bello'.

What Upsets Volcanoes?

Taking flight from a restrained expulsion
of Antuco volcano, Chile.[1]

'What is it that upsets the volcanoes
That spit fire, cold and rage?'
— Pablo Neruda, *The Book of Questions*[2]

A young geologist, Charles Darwin, was resting in woodlands in southern Chile, when, half an hour before noon, his peace was ruptured by a sudden rise and fall of the ground.[3] It lasted two minutes, shook the trees and left him giddy. He later confided in his diary that day, 20 February 1835, that such an earthquake 'at once destroys the oldest associations; the world, the very emblem of all that is solid, moves beneath our feet like a crust over a fluid'. Two weeks later, he rode into what remained of the city of Concepción, close to the epicentre of the tremor. The devastation aroused mixed feelings. It was 'the most awful yet interesting spectacle' he had ever beheld, 'terrible', yet, in its evocation of ancient Greek and Roman ruins, 'picturesque'. The scene at the port of Talcahuano, which had been ravaged by tsunamis, was even more desolate – a schooner had been tossed into the town centre – but at least the inhabitants there had reacted to the sight of a receding sea and found refuge on higher ground in time. Survivors blamed the earthquake on the curse of an indigenous Mapuche woman, rumoured to have stoppered Antuco volcano over a dispute. This tale intrigued Darwin as it mirrored his own views on the relationship between earthquakes and the suppressed activity of volcanoes. Plug a crater, he argued, and it will surely build up pressure elsewhere.

Darwin was further struck by the upheaval of land that had accompanied the earthquake. On Santa Maria island, just offshore, beds of recently thriving mussels had been thrust above the high-tide mark. Writing in his field notes, Darwin connected these phenomena – the earthquake, uplift, muzzled volcanoes. His ruminations led, on his return to England, to his first major published work, which identified one of the now-recognised triggers of volcanic eruptions, and came remarkably close to pre-empting a central plank of plate tectonic theory – the 'subduction zone' – by well over a century.[4]

Darwin made contact with many informants who shared their perspectives on the earthquake and its side-effects. One correspondent asserted that, at the instant of the shock, Osorno volcano, south of Concepción, vomited 'a thick column of dark blue smoke'. Lava had then 'boiled up' from a new crater, blasting out incandescent rock. The source also claimed that even the more distant Minchinmávida volcano had erupted as the ground trembled, and that nearby Corcovado volcano had followed suit before the year was out. Meanwhile, the British Governor on lonely Robinson Crusoe Island, far off the Chilean coast, reported an eruption at sea just hours after the earthquake.

The apparent synchroneity and geographic range of these phenomena astounded Darwin. Drawing the analogy of stamping on a frozen pool, forcing water up through dispersed holes in the ice, he imagined Chile floating on a 'lake of molten stone'. He reasoned that lava was evacuated through the volcanic orifices in the cordillera at the moment of the shock, and that repeated intrusions of liquefied rock were responsible for cumulative surface uplift. This explained how beds of fossilised seashells could come to lie at the top of a hill. Extrapolating in both time and space, Darwin thought that the whole of the Andes was subject to the same mysterious infernal force. He argued that the 1835 temblor marked just the latest 'step in the elevation of a mountain-chain' and that countless similar incremental upheavals might explain the formation of other mountain belts around the world.

While not all his facts were straight (for instance, there probably wasn't an eruption on Robinson Crusoe Island),[5] Darwin's pursuit of a unifying explanation for topography, seismicity and volcanism is extraordinary, the more so since his thinking diverged significantly from that of his then more established contemporaries. One prevailing view considered that mountain ranges were wrinkles in the skin of a cooling and contracting Earth. Today, geophysical observations have unveiled a fold in the Nazca plate where it collides with the South American plate – from Colombia to southern Chile – and sinks into the deep Earth under its own weight. This 'subduction' is the main motive force of tectonic plates, and it simultaneously acts as our planet's recycling factory by reprocessing old ocean floor and younger sediments. Subduction goes on to concentrate minerals in giant ore deposits, and fuel the world's most violent volcanoes with explosive magma. Masaya, Vesuvius and Stromboli are vivified in this manner.[6]

But the Andean plate boundary is not so well lubricated – it sticks, accumulating strain in the rock that is eventually released catastrophically by 'megathrust' temblors. For 135 years after 'Darwin's earthquake', this segment of the plate margin remained locked, even though the two sides continued grinding together at about the rate that hair grows. On 27 February 2010, the pent-up energy was unleashed in another massive earthquake. The towns of Concepción and Talcahuano were devastated once again, and 500 people lost their lives.

★

Understanding the triggers of volcanic eruptions is one of the keys to prediction. Darwin's image of the crust floating on a sea of molten rock is wide of the mark – magma bodies are very localised. But the idea that great earthquakes can provoke volcanic eruptions retains scientific currency and is backed up by statistical evidence. Although calculations suggest that the disturbance of seismic waves in a magma reservoir must be very subtle, in a region with many volcanoes, a few may be primed for eruption at any given time, and a nudge may be sufficient to unclamp the forces holding back the molten rock.[7]

There are at least sixty volcanoes in southern Chile, a high concentration for any country. Fifteen months after the 2010 earthquake, the volcano Cordón Caulle exploded violently – its largest historical eruption. Closer to the epicentre, Llaima volcano did not erupt but it did rumble.[8] More surprisingly, five of the region's other volcanoes subsided ever so slightly, possibly because the seismic rattling shifted geothermal waters trapped below ground.[9]

While Darwin's accounts and more recent events have furnished some of the strongest evidence for earthquake triggering of eruptions, the link is not yet clear enough to provide a reliable means of forecasting. In fact, Chilean volcanologists have often been taken by surprise by the country's eruptions. One of the most dramatic examples of this was the reawakening in May 2008 of Chaitén volcano, a small offshoot of snow-clad Minchinmávida. The initial fire and fury expelled dark towers of ash that scintillated with lightning. The plume soared into the stratosphere and fine ash drifted into air corridors, disrupting flights across the Southern Hemisphere. The fallout covered a wide swathe of southern Argentina, a nation that has repeatedly had cause to resent being situated downwind of Chile's volcanoes. Luckily, despite being completely unprepared, the 4,600 residents of Chaitén town at the foot of the volcano evacuated safely.[10]

It is even possible that the 1960 Valdivia earthquake, the largest ever recorded, lit a slow fuse for the Chaitén eruption. The seismic palpitations from that event appear to have opened a pathway for viscous magma stored deep under Minchinmávida to rise into a fault zone below Chaitén.[11] It gathered there over the next half a century until it became so pressurised that it popped the overlying crust like an overfilled balloon, squeezing molten rock up to the surface. Most volcanic eruptions are thought to arise in this fashion, from overinflation of a magma chamber. Sometimes, magmas with distinct chemical compositions are stirred together, producing pumice that looks a lot like marble cake.

Before Chaitén's spasm, Chilean volcanologists thought it had not erupted in more than 9,000 years, and ranked it fortieth in their list of volcanoes to worry about.[12] With limited resources to devote to surveillance, it did not make the priority list. Had it been monitored, precursory phenomena would have been detected. This is certain, because it later emerged that seismometers on some of the volcanoes higher up the pecking order *had* detected unusual activity beneath Chaitén.

Concerned by this forecasting failure, the Chilean president at the time, Michelle Bachelet, funded a new agency tasked with surveillance of Chile's volcanoes and identifying the signs of impending eruptions. I saw the fruits of this a decade later on a visit to the Volcano Observatory of the Southern Andes in Temuco. It is the nerve centre for sensors on volcanoes that stretch from the Altiplano, a vast volcanic plateau in the north, to Patagonia in the south. The main operations room was a large, airy and light space with three rows of desks facing dozens of large display panels affixed to the wall. Each screen showed live data feeds, mostly from seismometers but some from gas sensors, GPS equipment and cameras. The desks also bristled with computer panels. Personnel in the front row were tasked with reviewing all this real-time information. In the event of spotting an anomaly, they would alert staff behind them, who would then evaluate the signal in light of archived data. If it still stood out, the case was passed to the third tier for detailed contextual analysis. With so many volcanoes to track, I could see the sense in this hierarchy.

Even with this impressive facility, stealth eruptions still occur. My Chilean colleagues might cite Calbuco's 2015 eruption as an example. It gave only three hours' advance notice of its intentions via the observatory's seismometers, leaving very little time to react. Even more incredulous were the residents of Puerto Montt, as they watched a gigantic mushroom-headed plume of ash soar above the city skyline. Fortunately, there were no casualties, but farmland was extensively damaged, especially downwind in Argentina.[13]

A forensic chemical analysis and microscopic imaging of the pumice deposits in this case showed no evidence of magma mixing. Instead, it looks like the eruption was set in motion by a process known as 'second boiling'.[14] As magma cools in the Earth's crust, crystals form and grow, concentrating dissolved water in the liquid rock. Eventually, the magma can hold no more water, which then has no choice but to form vapour bubbles. A layer of foam accumulates, pressurising the magma chamber until a crack opens in the enveloping crust, sparking an eruption, which can be every bit as violent as that resulting from an overfilled magma chamber.

My PhD research was not directly concerned with identifying eruption triggers, but it did seek ways of identifying temperature anomalies that might signal the touchpaper had been lit. With so many volcanoes at different stages in their eruption cycles, it was inevitable I would carry out some of my field research in Chile. It was 1990, and the whole country was undergoing a different kind of seismic shift – Augusto Pinochet had relinquished power and a new government was about to be sworn in. In Santiago, a city of ice-cream parlours and noxious air, I watched President Aylwin's inaugural speech on TV, and sensed the air of optimism.

I soon discovered that democracy was not the only excitement, though. Láscar volcano, situated in the north of the country, had just exploded, expelling a plume of ash and steam into the stratosphere. Residents of the nearby communities of Talabre and Toconao were anxious, since a renowned geologist and Antarctic explorer based at the University of Chile, Oscar González-Ferrán, had urged their immediate evacuation. Half the mountain might be blown away with further eruptions, he claimed. At least, that was how the newspapers reported it. It hadn't been my intention to head into the heart of an evolving volcanic crisis but it was hard to ignore since Láscar was the target of my fieldwork, and my host was the Chilean geologist Moyra Gardeweg, who was suddenly tasked with responding to the emergency. As luck would have it, also in our party was one of the world's leading volcanologists, Steve Sparks. I had taken his lab classes (on the 1980 eruption of Mount St Helens) as an undergraduate. Whereas I turned to volcanoes because I'd never had a better idea, Steve must have had little choice with a name like Sparks.

Flying north from Santiago to Calama on the Tropic of Capricorn, you cannot take your eyes off the Altiplano, the largest, highest, driest plateau in the Americas. Perhaps only on Jupiter's volcanically hyperactive moon, Io, is creation and destruction by volcanic agency laid bare with more exuberance and strangeness. Almost devoid of vegetation, all the tints, tinctures and textures of weathered and varnished andesite, dacite and rhyolite are displayed: gravels and dusts in shades of rust, umber, ochre and maroon run together from ridges to valley floors. The inner Earth has been at play here, and these geogenic pigments have yielded sought-after resources.

We flew above Chuquicamata, the most prolific copper orebody in the world, spawned from subduction and magmatic activity dating back thirty million years. Before copper, it was the Atacama desert's vast reserves of saltpetre that enticed miners. This nitrate mineral was essential for making fertilisers and gunpowder. But the extractive industry collapsed with the discovery of a synthetic manufacturing process during the First World War, and the mines were abandoned, the shacks left to warp in the unforgiving sunlight.

Moyra had an appointment at the district government office in Calama to discuss the Láscar crisis, and Steve and I tagged along. A small crowd had gathered around a bearded guy playing the guitar in the town's plaza and singing '*Compañero*, here today, you are free . . .' The lamp posts were festooned with posters imploring 'Where are they?' and showing faces of Chile's 'disappeared', the people who had vanished under Pinochet. Some of them had ended their days not far from Calama at one of the saltpetre mines, repurposed as a concentration camp for political prisoners.[15] It features in Patricio Guzmán's stunning and harrowing film, *Nostalgia for the Light*. The dust of the disappeared blows in the desert's burning wind.

Attending the meeting were the new mayor for the region (appointed since the change of president), the provincial governor, the mayor of San Pedro de Atacama (the nearest town to Láscar), civil protection officials and members of the press. Provoked by Oscar González-Ferrán's widely publicised alarm, the army had stationed trucks in Talabre in readiness for an evacuation call. The governor asked if Moyra shared González-Ferrán's point of view. She demurred, explaining that without monitoring equipment, it was hard to read the volcano's affairs or foretell more violent eruptions. However, in light of Láscar's past behaviour, she viewed the recent explosion as typical of the life cycle of the dome of viscous lava known to reside on the crater floor. Any major escalation in activity, she argued, would likely be preceded by ground shaking strong enough to be felt in villages around the volcano. She added that the cost of a seismograph, which could give early warning, was much less than that of resettling Talabre's inhabitants.

Scenting controversy, the journalists jumped on the apparent divergence from González-Ferrán's conviction. But Moyra would not contradict a colleague's assessment. In an equal display of diplomacy, the head of the regional emergency office noted that González-Ferrán was a two-hour flight away in Santiago; accordingly, he would listen to the expert who was present. The army would remain on standby and seismic equipment would be obtained. But, he wanted to know, if evacuation should prove necessary, how long might it last, and what should be done for the thousands of livestock, mostly llamas?

For a rookie volcanologist as I was then, the exchange was gripping – it revealed the complexities and conflicts of managing a volcanic crisis, and the importance of science in informing critical decisions affecting people's lives and livelihoods. The area around Láscar is sparsely populated; fewer than 100 people inhabited Talabre. While traumatic for those impacted, an evacuation order would not have presented a major logistical challenge. But transpose such a crisis to an urban area, and the anxiety surrounding decision-making amplifies intensely.

An hour or so's drive brought us to San Pedro de Atacama, where we spent the night. Along with the scents of weeping eucalypts in the plaza, kitchen woodsmoke and mule droppings, I couldn't help noticing a sulphurous reek, always rousing for a volcanologist. It didn't take long to find out that sulphur was being mined from the 5,500-metre-high summit of nearby Purico volcano, and refined at a plant on the edge of town.[16] The mine has since closed, but the volcano still has its uses as a lofty foundation for the Atacama Cosmology Telescope.

In the morning, Moyra, Steve and I set off in two four-wheel-drive vehicles barrelling south down the undeviating 'Road of Patience', crossing the eastern edge of the immense Salar de Atacama salt flat. The latest extractive industry to flourish here is the harvesting of lithium, sourced from magmatic fluids and weathering of volcanic rocks, and distilled in the Salar's brines. As the power source of rechargeable electric batteries, demand for the metal has rocketed. The abundance of lithium in the region is even evident in the hair of ancient Atacamanian mummies.[17]

After so much anticipation, first impressions of the Andean volcanoes from the ground were underwhelming. The cones formed indistinct silhouettes, far less imposing than I had imagined; the scenery seemed, if anything, drab. But I was being tricked by a collusion of atmosphere and terrain – what seemed a few miles away was actually ten times more distant. Adapting to this illusory perspective, my excitement rebounded as we turned east and began ascending the Altiplano's massive piedmont.

This rampart was forged during the most violent volcanic cataclysms the Earth has known: sustained explosive affairs so extreme that we have no other terms for them than 'super-eruption' and 'mega-colossal'. Super-eruptions produce at least one trillion tonnes of rock, known as ignimbrite. Recent analysis of the geological record suggests such gargantuan events recur every 20,000 years or so, more often than once thought.[18] Whereas most eruptions are triggered when magma chambers reach a critical pressure due to overfilling or a build-up of bubbles, super-eruptions may result when the sheer buoyancy of an immense reservoir of molten rock fractures the lid of overlying crust – like releasing a football held underwater.[19]

Ignimbrite is a rock made up of the ash and pumice deposited from tumultuous plumes of magmatic debris and gas that have surged, with the combined qualities of hurricane, landslide and furnace, from giant rents in the Earth's crust. Ignimbrite can be powdery and friable, but where it has sintered through stewing in its own juices soon after eruption it forms a resilient lightweight rock known as *sillar*, well suited for construction. Its desert varnish has an orange-pink tinge, but a fresh surface is dazzlingly bright in the intense Atacama sunlight. Super-eruptions expel so much magma that the ground founders above the voiding magma chamber, forming a vast depression. These are not the classical volcanic cones we are familiar with; they are, instead, calderas – exceptionally large holes in the ground.

Moyra was particularly familiar with the ignimbrite we were driving up; she had examined it in detail and written the bench-mark paper on its age and characteristics.[20] It is linked to a massive caldera the size of Greater London, known as La Pacana, and is nearly four million years old. If spread evenly across the entire USA, the debris would reach above ankle height. For comparison, Yellowstone's greatest convulsion is half La Pacana's age, but the deposit is similar in scale.

The Altiplano is a vast palimpsest of such ignimbrites, the products of innumerable super-eruptions. It holds aloft the discrete cones of active and dormant volcanoes whose mouths touch attenuated air at up to 6,800 metres above sea level (higher than Mount Denali in Alaska). Clefts in the ignimbrite deep enough to hide an apartment building, known as *quebradas*, give a sense of the scale of individual eruptions. Arriving in Talabre village, where we would be staying for the next two weeks, the telluric became domestic: the tin-roofed dwellings are all built from sillar. I discovered that Talabre had already been relocated further from the volcano in the 1980s owing to the threat of mudflows, but farmers still tended maize and vegetable gardens around the former settlement. They also kept herds of llamas for wool, which was woven into the rugs I'd seen hanging in San Pedro's tourist shops.

We stayed at the village schoolhouse, laying out sleeping bags on the floor of the classroom, which we shared with soldiers on standby to assist should an evacuation order be issued. A picture of the Pope hung from the wall, and beside it an empty frame that had until recently ornamented Pinochet's portrait. The air was frigid and gusty; the Chilean flag on the roof clattered all night. In the morning, the ground was icy, and white steam clouds wafted from Láscar's snow-flecked summit. The school's teacher, Manuel, was originally from Santiago. He was plainly disaffected by his posting to such a remote and forlorn frontier, all the more so now it was menaced by an unruly volcano. He told us the recent blast had rattled the window for five minutes.

One of Talabre's oldest residents, Cornelius, shared a longer-term perspective. He recalled many eruptions since the 1950s, including one in 1960 that had continued for three days, showering the village with ash at night. That had been preceded by two months of ground shaking, he said, and since Láscar was now tranquil, he reckoned it would stop smoking soon and go back to sleep. This prediction was strikingly in accord with Moyra's interpretation – pathways by which gas escapes from the magma are known to clog up with mineral growth, shutting down gas release. Pressure then accumulates to the point of a violent expulsion. Cornelius even described the present condition of the volcano as an 'open crater', consonant with the term 'open vent' used by volcanologists to describe exposed magma that relieves its internal pressure by unrestrained emission of gas into the atmosphere.

While unconcerned about the volcano, the prospect of evacuation of the village troubled Cornelius. 'Water is our most precious commodity here – it comes from the spirits in the volcanoes,' he explained. 'Someone might take it if we are forced to leave. And if we move again, we will have to dig a new channel to carry the water, and this costs us in money and effort.' Echoing the feelings of the octogenarian I met on Stromboli, Cornelius described the volcano as a friend and protector. 'Láscar announces the change of seasons by the clouds at its peak; it tells us if the year will be good or bad, and watches after our animals.' Cornelius had heard from various authorities in the preceding days that 'the geologists will know when the volcano will erupt', but in truth, without any monitoring equipment, his direct familiarity with the volcano trumped our generic scientific experience.

Andean volcanoes have long been revered as sources of water, providers of fertility, guardians of animals and owners of minerals and obsidian, a volcanic glass used to make cutting tools. Cornelius's faith in the supernature of Láscar echoes ancient beliefs. The most striking (and lurid) evidence of this dates to the Inca period (circa 1438–1532 CE), when human sacrifices (now more neutrally referred to as 'deliberate installations')[21] were performed beside fuming craters. The chronicles of the Spanish conquistadors record that the most handsome children from the span of the empire were brought to Cusco, where they were fattened up and prepared for mountaintop immolations known as *capacochas*.[22] Spectacular relics substantiating these accounts have been unearthed, including mummified victims from what must be the world's highest archaeological site at the summit of Llullaillaco volcano, to the south of Láscar. In 1999, three burial pits were discovered within a rough ceremonial platform just below the peak and more than 6,700 metres above sea level. A boy and girl, both aged four or five, and an older girl of about thirteen, were exhumed. Though cause of death could not be ascertained, each child had been heavily sedated with maize beer. It is possible they were buried alive. Their clothed and hunched bodies were accompanied by jewellery, ceramic and wooden pots and plates, foodstuffs, and gold and silver figurines of llamas.

What led the Inca to perform such extreme rites is unclear. They worshipped the sun, which the craters brought them closer to, linking heaven, earth and underworld, but in propitiating local deities, it appears the *capacochas* were primarily intended to reinforce Cusco's authority over far-flung parts of the vast Inca dominion. Given its altitude, sulphurous essence and proximity to the Inca road, Láscar's peak may well have been another site of 'deliberate installations', though any physical evidence would surely have been obliterated long ago by the volcano's fiery numen.

We spent several days geologising around Láscar. As a novice, especially in the field identification of young volcanic rocks, I experienced a kind of rapture in the company of Moyra and Steve, who is also an ignimbrite guru. They combed each outcrop for clues to interpret the emplacement or sedimentation of the various lavas and pyroclastics (meaning 'broken by fire' and referring to the granular mixtures of ash and pumice generated by explosive eruptions). Steve routinely thought aloud, and would illustrate schematics of underlying physical processes in his notebook. I quickly recognised two central and related aspects of volcanology. First, understanding volcanoes means understanding fluids – the same physics applies to anything that flows, whether water from a tap, oil in a pipe, blood in our veins, or magma, lava, mud and ash clouds moving below, across or above the ground. Second, the colour, thickness, lamination and texture of the layers of pyroclastic rocks and lavas faithfully record the nature of the eruptions that produced them: how high the ash cloud reached in the atmosphere; how intense the explosions were; how long they continued for; which way the wind was blowing. All this can be deciphered for any past eruption given decent exposure of the deposits and knowledge of how things flow. If you know how to read the signs, it is all written in the rocks.

The road climbs from Talabre to a dizzying 4,500 metres altitude, winding around the southern foot of Láscar. Here, tussock grass – resembling oversized porcupines and equally uncomfortable to sit upon – grows in profusion, adapted to the hyperaridity, and complementing the burnished patina of cinereal and russet lava rock of older volcanoes. Via this route, which leads to the Argentine frontier, we accessed the east side of Láscar, an astonishing landscape and ecosystem at the southern margin of La Pacana caldera. We sighted many rheas and vicuña (wild cousins of domesticated alpaca), while vizcachas (leporine rodents) watched us inquisitively from behind blocks of lava. I imagined the ancient Atacamanians hunting in this place.[23]

Láscar loomed beyond a dark green lake, Laguna Lejía, whose motionless surface perfectly reflected the volcano, now shrouded in iridescent lenticulars. Pink and white flamingos shimmered beside the salt-fringed shore before taking flight, bill to tail, above the water. The stony plain stretching north of the lake was pockmarked by the hail of projectiles from the recent eruption. Each hollow contained grey fragments of a shattered lava bomb.

Since hardly anyone ever looked directly into the crater of Láscar, some of the most detailed insights into its activity had come from satellites in space.[24] In particular, the sensors on the U.S. Landsat orbiters had kept an eye on the volcano. The Landsat programme, operational since 1972, was the first to bring into the public domain the kinds of Earth imagery we are now so familiar with from online digital globes. I was particularly interested in the satellites' capability of detecting very hot ground but I wanted an image collected at night so that only volcanic glow would light up the pixels – by daytime, sunshine reflected off the ground would swamp the signal. Since there wasn't much of a market for views of the Earth in darkness, I'd had to make a special request for Landsat's data recorder to be switched on at the right moment. My plan was to be at the crater's edge by dusk, shortly before the satellite would whizz overhead.

At noon that day, Steve dropped me off where the road passed closest to the volcano. In my backpack were my trusted infrared thermometers, veterans of the Stromboli campaign. Steve would return for me at 10pm.

In terms of terrain, there was nothing challenging or technical to the climb. There was no route or path to follow, and though this was before the era of commercial GPS, it was obvious enough that the destination lay upwards. However, starting at around 3,800 metres above sea level, and unacclimatised, the thin air throttled my vitality. By 4,500 metres, I could manage two steps up before having to catch my breath, during which time I would slide back in the loose cinders. It was gruelling, heart-pounding work, and a bothersome headache was brewing. As the afternoon wore on, the volcanoes cast great shadows across the plateau, until just the snow-dusted peaks glowed cherry-blossom pink against a silvery sky.

At last, around dusk, I approached the crater's edge at close to 5,400 metres altitude. But there was little respite, as what the air lacked in oxygen it made up for with sulphur emanating from the pit. Feeling sick, I peered into the great maw of the crater and, through the fumes swirling within, discerned a constellation of red-hot glowing gas vents arranged in clusters and arcs on and around a circular pile of lava, about the size of a baseball field. The lava surface looked wrinkled, like an underbaked cake that has sagged. Some vents beyond its margin roared like jet engines. By now, I was not on top form and was shivering in the sub-zero air, but it was thrilling to know that, hurtling in orbit above me, Landsat was about to be activated to capture the scene. It would include a sub-pixel sized human who knew exactly what the volcano was up to at that moment.[25]

Keeping an eye on time, I hurriedly took temperature readings, notes and photographs. I did not want to miss my appointment with Steve. Unfortunately though, as I began the descent, my torch batteries expired (maybe they'd got too cold), and I had to find the way down in the dark. While ascending a volcano leads you to a singularity in space, descending does not: it was essential to retrace my steps or risk ending up at the foot of the mountain miles from the roadhead – this was not easy by the light of the stars alone. My legs ached; I was exhausted and nauseous, and soon dealing with diarrhoea and vomiting, all clear signs of acute mountain sickness. I felt an overwhelming urge to lie down and rest despite the cold, but knew I must continue or risk raising the alarm and having a search party out for me in the morning.

Several times I reached the head of a gully and was unsure of which side to descend. After more than three hours, I reached the Talabre *quebrada*, somehow only a quarter of an hour late for the rendezvous. But I could not see any vehicle. I was about to sink to my knees when full beams flashed on, and Pink Floyd's *The Dark Side of the Moon* unexpectedly resounded across the valley.

I spent another week in Talabre and climbed Láscar twice more, but from the east and with absolute ease thanks to adapted blood chemistry. One morning, an elderly woman stopped by the schoolhouse and shared her opinions on the crisis. 'The volcano did a little thing,' she said, 'like it has done many times before. Now it is all quiet.' Moyra had reached the same conclusion and recommended calling off the alert. The soldiers retreated to San Pedro. Only Manuel, the teacher, looked sorry – he had privately hoped for an evacuation to bring an early end to his lonesome secondment. He was given charge of a seismograph, though, and trained in its maintenance. I like to imagine that through listening to Láscar's subterranean whispers, he came to find a friend in the volcano.

★

Back in the UK, I eagerly awaited delivery of the Landsat images. Since I was interested in developing ways to apply the satellite remote-sensing techniques to routine volcano monitoring, I obtained more than a dozen additional images of the volcano spanning a seven-year period. The temperatures and hotspot sizes calculated from the scene acquired the evening I spent at the summit were reassuringly consistent with my field observations. But most interesting, the trends that emerged from my analysis of the data correlated with eruptive activity. Láscar's largest blast was preceded, somewhat counterintuitively at first glance, by two years of steady cooling. This can be explained by the sealing of cracks and fractures in the lava and surrounding host rock, cutting off gas flow, pretty much as Cornelius surmised. In this scenario, pressure in the magma conduit escalates until a threshold for rock fracture is reached, triggering an explosion. Blasting off the carapace of the lava dome depressurises the underlying magma, kindling a runaway process due to violent expansion of gas bubbles in the molten rock. The fresh magma rising up in the waning stage of the explosion cools, sealing up the conduit, acting like a valve near the top of the volcano, and resetting the cycle.

More recently, a connection has been drawn between powerful steam blasts, known as phreatic eruptions, and rare rain and snowfall events on Láscar – in other words, weather can trigger eruptions. This was recognised in 2015, when rainwater infiltrated the lava dome and came into contact with hot rock, generating steam and rock blasts like a faulty pressure cooker blowing its lid.[26] Such activity, recognised at volcanoes worldwide, typically poses no threat to communities beyond the flanks of the volcano, but the consequences can be tragic for tourists (and volcanologists) who are in the wrong place at the wrong time.[27] This possibility is often on my mind when I do fieldwork on active volcanoes. Until we find reliable ways to forecast such relatively weak events, the simplest way to reduce the risk of getting hurt is to limit time spent in hazardous areas.

Some of the methods I developed during my PhD studies are still in use and being applied to routine monitoring of Láscar. A study in 2020 presented an analysis of hundreds of Landsat images of the volcano spanning almost four decades, revealing with much greater clarity its cycles of lava-dome construction and explosive destruction, and corroborating the idea that sealing and pressurisation of the lava dome initiates eruptions.[28]

Landsat 9 was successfully launched in 2021, a year before the programme's fiftieth anniversary. No other satellite remote-sensing system comes close to providing such continuity and consistency. The breadth of Earth observation and change detection studies it has tackled is equally unrivalled. For scientists devoted to synopsis, it is a formidable ally. There have even been calls for a dedicated orbiting volcano observatory capable of monitoring the world's volcanoes and detecting when their fuses are lit.[29] The Inca looked up to the heavens from Andean volcanoes; now we watch the lofty peaks from above. I still believe, though, that the spaceborne imagers, as well as the robots, drones and machines, have got a way to go before they eclipse a trained field volcanologist huffing and puffing up and down the mountain on foot, with cases of sensors and samplers strapped to her back, in search of 'ground-truth'.

Emerald Isle

Mr Frank A. Perret listening to the subterranean
road of a volcano at Pozzuoli, Italy.[1]

*'I look up at the mountain. She is still there. Disturbing the
leaves, the trees, the flowers, the sea, the animals, and all human
kind.'*

—Yvonne Weekes, *Volcano: A Memoir* [2]

Friday 9 May, 1902. One can only imagine the dread that must have overcome Mr Outerbridge, owner of the Quebec Steamship Company, as he read the following cable:

WE NOTIFY YOU THAT STEAMSHIP RORAIMA HAS BEEN LOST. WE DO NOT KNOW WHETHER PASSENGERS WERE ABOARD OR NOT. VOLCANIC ERUPTION. LOSS TERRIFIC.

What befell the vessel emerged later from the traumatic testimony of Edward Freeman, captain of another ship, the *Roddam*. Freeman had dropped anchor near the *Roraima*, on the morning of 8 May, in the sweeping bay of St Pierre, the vivacious commercial and cultural hub of Martinique island. Landmarks of this jewel of the French colonies included a court of justice, a cathedral, a grand theatre and botanical gardens. On this occasion, Freeman was struck by the sight of Montagne Pelée, the normally peaceful volcano that overlooked the city: 'The smoke came pouring out like big rolls of wool from all sides . . . curling up in the shape of great cauliflowers.'[3] But despite a light fallout of white ash that dusted the ship, the sun still shone on the pretty red-roofed houses and sugar estates, and on the other steamers and sailing boats at anchor. Church bells were pealing for early Mass – it was Ascension Day.

Then, all of a sudden, Pelée sounded a shattering blast and unleashed a 'column of flashing flame'. At first, Freeman watched spellbound, but then with mounting horror as a 'black pillar of cloud' raced down the flank of the volcano like a tornado. Within two minutes, it had cleared the foothills, fanned out and swallowed St Pierre from north to south. Freeman barely had time to question the fate of the 27,000 townsfolk before the raging cloud lashed the sea, triggering waves that knocked several boats straight to the bottom. He hurled himself into the chartroom just before the *Roddam* was struck and nearly capsized. Those on deck were thrown overboard and drowned in a 'boiling sea'.

But Freeman could not escape the terrible cloud. As it enveloped him, it felt as if handfuls of red-hot dust were being pressed into his face and hands, but worse still, each breath filled his lungs with scorching ash. Suffocating, burned inside and out, the ship still jostled by violent seiches, he managed to grope his way to the bridge, tripping over bodies. Searing ash plastered the deck and ignited fires. Lifeboats and rigging were aflame; dying men shrieked and groaned in agony; some ran around 'like raving maniacs', while others jumped into the water in forlorn hope of escaping the 'blazing, blinding shower'. Using his elbows on the telegraph handle, Freeman signalled 'full speed astern'. Spotting the chief engineer, he asked if the anchor was fast, but the officer was incapacitated and soon succumbed to his injuries.

Only the assistant engineers, who had been in the engine room, were unharmed. Freeman mustered a crew of walking wounded and attempted to steam out of the bay, playing the anchor out until it snapped. But the ship lurched like a 'wounded monster', with the starboard steering gear jammed with wreckage and pumice. Waves kept rolling them to shore, and at one point they almost collided with the *Roraima*, blazing furiously at her stern while smoke billowed from below. Survivors crowded her forward decks, but it was impossible to help them.

At last, Freeman gained control of the ship, and aimed for open water. Breathing was excruciating. A crew member spent the next hours wiping the captain's scorched and weeping eyes so he could navigate.

At last, the smouldering vessel, 'blistered and baked to a crisp' and coated in what looked like bluish-grey cement, limped into Castries harbour on St Lucia. One onlooker recalled wondering, with astonishment, how it was possible 'for a ship to get into such a terrible condition and then float'.[4]

Freeman recovered in hospital. His recollections provide a rare and vivid account of immersion in volcanism's most pitiless manifestation, *nuées ardentes*, or 'burning clouds'. The fate of St Pierre and its 27,000 victims is further memorialised in the name given to this variety of eruption, 'peléan'.[5]

It is said that the only survivor in the direct path of the *nuées* was a prisoner, Cyparis, held in solitary confinement for bad behaviour. Even in his bunker-like cell, he sustained dreadful burns. A few of the *Roraima's* crew and passengers were rescued when a French cruiser, the *Suchet*, arrived on the scene. But there was no sign of the captain. Like many others, he had jumped into the sea. His body was never recovered. Sharks saw to that.

Martinique's tragedy was front-page news across the world. Some geologists noted the parallel destinies of St Pierre and Pompeii, not realising how germane the comparison was, since there was then no scientific understanding of *nuées ardentes*.[6] For a thirty-four-year-old engineer, convalescing at home in Brooklyn after a period of depression, empathy for the victims behind the grim newspaper headlines, and the scientific bewilderment concerning what exactly the volcano had done, sparked an extraordinary career change. Formerly a designer of elevators, fans and pumps, Frank Alvord Perret became the foremost volcanologist of his generation, observed peléan eruptions from the verge of *nuées ardentes*, and foretold the enduring volcanic crisis that would befall Martinique's neighbour, Montserrat, half a century after his death.[7]

Perret's volcanic epiphany coincided with his physician's advice to turn his back on the past and rehabilitate somewhere warm, fragrant and diverting. Somewhere like Naples, the doctor offered. And so Perret followed in Sir William Hamilton's footsteps, becoming as infatuated with Vesuvius as had the British diplomat and collector. It did not take long for Perret to befriend the volcano observatory's director, Raffaele Matteucci, who in turn realised the American's scientific acumen and ingenuity might be put to good use. Matteucci, an incurable volcano-lover who lived 'in mysterious and terrible solitude . . . bound together forever' with Vesuvius,[8] also saw a kindred spirit in Perret, another man ready to devote his life to lava.

Perret thought of Vesuvius as the perfect 'cabinet volcano' on account of its 'small size, accessibility . . . diversity of eruptive phenomena [and] rich mineralogy'.[9] Just an hour's walk from the observatory, the hissing crater was a natural laboratory. Over the next two years, he honed his skills of volcano observation, using a camera, sketchbook and simple devices of his own construction to record earthquakes, gases and sounds.[10] He relied, too, on his senses. A well-trained nose, he asserted, could analyse a range of volcanic gases, including sulphur dioxide and hydrogen chloride. In March 1906, sensing a constant buzzing from below, he bit the metal frame of his bed to amplify the tremors with his skull. It told him change was afoot. Not much later, Perret spied the merest hint of ash in the steam plume emanating from the volcano. No one else would have noticed, but for him it was 'the signature of a new power'. Sure enough, by nightfall, a fissure had opened near the summit, spewing dark clouds that showered Naples with ash. The eruption intensified over the following days; lava inundated towns on the south-east flank of the volcano, while communities to the north-east bore the brunt of ash fallout. In the town of San Giuseppe, more than 100 people sheltering overnight in the church were killed when the roof collapsed under the weight of ash. In all, more than 200 lives were lost, and the economic losses crippled Italy.

Throughout the eruption, Matteucci and Perret faithfully remained at the observatory, which was perched on a hill not far from the crater. Here were two utterly absorbed observers – so attuned to the volcano through their instruments and senses that no variation in the activity or possible connection between phenomena escaped their attention. They recorded events and sent telegrams to the authorities with advice on evacuations.[11] In the small hours of 8 April, the seismic shocks became so strong they left the building and watched and listened in awe as a towering column of fire roared high above. Bolts of lightning discharged to earth from it and 'fire avalanches' of incandescent lava bombs slid down the steep cone. The men pulled overcoats across their heads to protect them from the hail of red-hot stones. Perret recognised an 'infinite dignity' in the sublime performance. 'Each rapid impulse', he wrote, 'was the crest of something deep and powerful and uniform which bore it, and the unhurried modulation of its rhythmic beats sets this eruption in the rank of things which are mighty, grave and great'.[12]

Perret continued to work on Italian volcanoes, studying Stromboli during an eruptive crisis in 1907 (when he dissuaded the navy from an unnecessary evacuation of the population), and Etna in 1912 (where he was badly burned and bruised by volcanic projectiles). By then, his colleague Matteucci had died, his lungs inflamed from inhalation of so much volcanic ash.

Eager to expand his knowledge of the variety of volcanic activity, Perret went on to work in Japan, Hawai'i and Tenerife. On the Big Island of Hawai'i, he helped to establish the first U.S. volcano observatory. Its inaugural director, Thomas Jaggar, wrote later: 'I knew at once [Perret] was the world's greatest volcanologist. His skill was taking pictures. Mine was making experiments . . . these two skills in action would accomplish what theories never could approach.'[13] Perret did far more than operate a camera, though. To understand how volcanoes work, it was 'absolutely necessary to be at the crater at the time of eruption'.[14] Accordingly, Perret would construct huts or shelters as close to the action as possible. Then he would set up an array of monitoring equipment and fill notebooks with his own corporeal and sensory 'measurements'. He mostly worked alone, and likened his skills to those of a physician. As his experience grew, he became convinced that science could forestall volcanic disasters. But he took increasing risks: he was often too close to his patient.

Inevitably perhaps, Perret would confront the volcano whose calamitous eruption had recast his destiny: Montagne Pelée. He was in Puerto Rico when it reignited in 1929. He gathered his instruments and made hastily for 'the very theatre of the frightful catastrophe of 1902'. In no time, with help from the men, women and children who had rebuilt St Pierre, he had erected a hut close to the summit and overlooking the treacherous valley that had funnelled the fateful glowing clouds or *nuées ardentes*.[15] 'Excursion, observation, photography and intermittent sleep were the observer's lot,' wrote Perret. His investigation was 'as far as possible, *lived* on the mountain itself, day and night . . . amid penetrating ash, incessant noise and pungent smells'.

From his outpost, Perret observed hundreds of *nuées ardentes*. He even walked into a very minor example to ascertain its temperature and odour. He interpreted them as hot avalanches, explosively generated by pressurised gases that frothed the lava. It was the thorough dispersion of energy within the moving mass of rock, dust and gas, he argued, that enabled them to shapeshift and move with ominous silence.

Since we can only ever see the exterior of *nuées ardentes*, the deposits they leave and the damage they do, there remain gaps in knowledge of their physics to this day. Perret's deductions, made nearly a century ago, remain coherent with modern thinking, though we now refer to 'pyroclastic flows' and 'surges', or 'block and ash flows' rather than the more lyrical *nuées ardentes*. They are rather typical of subduction zone volcanoes, like Pelée and its neighbours in the Caribbean, and several processes can generate them, including the collapse of perched lava domes, or violent explosions whose plumes of rock and gas are too dense to stay aloft and so plummet back to earth. In both these cases, topography then plays the decisive role in channelling the searing hurricanes of gas, ash, pumice and blocks of lava down the volcano. Speeds can exceed thirty miles per hour, and pillars of hot dust shoot up, ingesting and heating air (a point recognised by Perret), to form cauliflower clouds. As the pyroclastic mixture travels downslope, larger rocks gather at the base, forming a dense 'block and ash' flow that traces the valley floor, fanning out where it hits flatter ground. The dilute mix of ash, pumice and hot air and gases above can be much more mobile, capable even of jumping from one valley to the next. Turbulent and dilute, or dense and confined, *nuées ardentes* are extremely destructive owing to their heat and force. What isn't crushed or asphyxiated may yet go up in flames.

One evening, Perret had a very narrow escape at his field station. Alerted by a strange noise, he rushed to the cabin door as a towering spine of rock on the dome crumpled, issuing a *nueé* of 'inky blackness'. Shortly, an explosion propelled a second avalanche and an accompanying cloud 'white as snow'. The two columns were hurtling towards him; there was no time to flee. He calculated he might be dead within four minutes. He wrote later of 'a sense of utter isolation; awe in the face of overwhelming forces of nature so indifferent to my feeble self'. As he darted inside the hut, the blast hit with 'swirling gusts of ash-laden wind'. Despite his immersion, Perret continued observing,

applying his trained nose to gas detection. It may be questionable how reliable sense of smell is when one is suffocating in ash
clouds, but Perret claimed the *nueé* was odourless, and he
excluded the presence of any acid gases. The 'spectacular
onrush' was also mute: *nuées* look tumultuous, clamorous, terrifying, but they are deathly quiet. There was no questioning the
unpleasant physiological effects, however; Perret's airways were
scorched, and he was left short of breath for weeks. The doctors
injected him with horse serum, a cure that sounds as unpleasant
as the affliction.

★

As well as a profound physical connection with volcanoes, Perret
was strongly motivated by humanitarian instincts. Like his
friend Thomas Jaggar, he thought of volcanoes as living organisms, and was convinced that through systematic observations,
eruptions could be predicted, and death and disaster avoided.
As an internationally renowned volcanologist, he was now often
called upon to opine on unfolding volcanic crises.

It was while on Martinique in April 1934 that Perret received
a telegram from Henry Meagher, the British consul on the
island. Meagher was enquiring about Pelée, but added enigmat-
ically 'Montserrat asking'. He was referring to the neighbouring
island, which was under British administration, and had a
volcano of its own. Perret followed up and discovered that
earthquakes had been rattling Montserrat, whose capital,
Plymouth, was even closer to the top of Soufrière Hills volcano
than St Pierre was to Pelée. The vigour of the hot springs and
sulphurous vents that gave the Soufrière Hills their name had
also intensified. Similar increases in activity had preceded the
1902 eruption of Pelée, knowledge of which understandably
unsettled the people of Montserrat. Perret arrived in Plymouth

the following month and felt at once that the situation 'might become serious'. There were those who saw the famed volcanologist's very appearance on the island as a sign of impending doom.

Perret did not stop long, though. He had to go to New York to assemble the equipment – vacuum tubes, thermometers, cameras, binoculars, gas syringes, pressure gauges and pendulums (to detect earthquakes) – that he would need to formulate more than a hunch about Montserrat's stirrings. Back on the island, he quickly established a field station near a zone of sputtering vents and boiling pools that mineralised into milky and inky streams. The tremors continued, some damaging buildings, while the fumarolic emanations grew more insufferable downwind.

Volcanic activity generates acoustic as well as seismic energy. Perret placed microphones in contact with the ground and listened to the volcano's pulse through earphones. He also installed air-reception microphones and tracked a glissando of gas vents at the *soufrières* – one that roared a B flat in December was trilling F sharp by March. With all this change and excitement, it annoyed Perret that he had to sleep sometimes – he might miss a significant shift in the activity. And though he had trained local observers to assist in documenting the activity of the volcano, they would instinctively take flight during an earthquake rather than write careful notes on the vibrations of the seismoscope. Clearly, round-the-clock surveillance demanded automation. Accordingly, Perret improvised a clockwork system to expose filter papers sensitive to changing gas concentrations, and designed a recorder for his microphones. He even grew sensitive plants whose leaves drooped when disturbed, and kept caged parrots to see if they might detect earthquake precursors, but neither flora nor fauna proved to have foreknowledge of volcanic activity.[16]

Other patterns emerged in his data, though, and Perret was convinced that the complex rhythms of volcanoes arose from subterranean and astronomical forces acting in concert. He noted, for instance, that seismic and volcanic crises in the Caribbean developed at roughly thirty-year intervals and linked shifting tidal forces (tuned by the positions of Earth, moon and sun) to the cadences of eruptions.[17] He was certainly on the right track – there is evidence for all manner of cycles in volcanic behaviour – but his explanations have not been verified nor have more general causes been identified.[18]

One night, a new gas vent opened with an explosion followed by wailing and whistling. Perret's thermometers were so corroded by acid gases they fell apart. He suffered severe nose-bleeds and his salivary glands swelled to the size of eggs, prompt-ing him to remark that 'volcanologists require a great deal of their instruments, even as they do of themselves'. But despite the escalating unrest, especially a terrifying tremor on 12 December 1934, Perret announced publicly that an eruption was unlikely. There wasn't enough evidence for him to reach clear conclusions, and he did not want to fuel panic unnecessar-ily. Instead, he shared his concerns privately with the governor, advising that plans be drawn up for a rapid evacuation of Plymouth, since an eruption on so small an island 'might well be cataclysmic', engulfing the whole place in minutes.

The crisis rumbled on through 1935. One earthquake, on 10 November that year, triggered landslides and demolished a church. Another left Perret with an 'indescribable feeling of there being an immense power behind it', an echo of Darwin's emotions on sensing the Chilean earthquake a century earlier. But more importantly, he realised that these shocks had deeper origins, signalling that the volcanic threat at the surface was waning. Sure enough, the unrest subsided. It had all been a 'subvolcanic' affair: rising magma had provoked the tremors and reheated the *soufrières*, but had failed to open conduits to the surface. Their rehearsal was over – the Soufrière Hills went back to sleep.

★

Perret's identification of a thirty-year cycle in the Caribbean has been remarkably borne out on Montserrat. The 1933–1935 crisis he studied so closely had been preceded by a similar affair in 1897–1899. Another came in 1966–1967, and yet one more in 1992–1995. It was this last episode that culminated in magma reaching the surface. Geophysicists at the University of the West Indies described it as 'the most long-expected and clearly-signalled eruption [of] the twentieth century'.[19]

You might think that this level of anticipation would have prepared the people of Montserrat for what seemed an inevitability. Sadly though, the authorities and public on the island were not ready to act when the Soufrière Hills shifted gear from 'subvolcanic' to volcanic in July of 1995. The scale of disruption, dislocation and loss of life that followed was, arguably, avoidable.

In fact, a detailed report on Montserrat's volcano had been presented to the island's government eight years earlier.[20] The report enumerated scenarios for future eruptions, indicating where devastating *nuées ardentes* would likely travel. It also highlighted the extreme vulnerability of Plymouth, situated in a valley that led straight to the volcano's craggy summit. But it was ignored or misunderstood, perhaps even lost. Some said it was washed into the sea when hurricane Hugo ravaged the island in 1989, leaving ninety per cent of the population homeless.

Allegedly, the three people most responsible for disaster preparedness on the island at the time – the governor, chief minister and head of police – had not even realised they were living on a volcano. Everything envisaged in the report became real.

Before the eruption, Montserrat's moniker was 'the Emerald Isle'. It celebrated its Irish heritage, gave us Arrow's bubbly (and dissonantly prophetic) calypso hit, 'Hot Hot Hot', and was the holiday destination of choice for those seeking calm. Windsurfers, parasailers and speedboaters went elsewhere. Many 'snowbirds' from North America built homes on the island, where they would spend the winter. In contrast, visitors today come to see the 'Pompeii of the Caribbean'. The emerald is caked in grey grit.

From the first powerful steam explosions that showered Plymouth with ash on 18 July 1995, Montserrat was set for irrevocable change. While many Montserratians fled the island, a host of development and aid organisation managers, contractors and consultants, as well as social scientists, film crews and volcanologists converged there. The eruption had been going on for a year when I joined the mêlée.

It was a sticky tropical evening when I arrived, and too dark to make out where the volcano was or gauge the lie of the land. I went straight to the observatory, but struggled to hear anyone above a hubbub of whistling frogs and a high-pitched portamento coming from a bank of seismographs frenetically transcribing, with ink pens, messages from seething magma. Over the preceding months, lava had been extruding gradually at the summit, building a gigantic dome of shifting rock, but it had just begun growing at a much more alarming rate. This mound was buttressed on three sides by toothy remnants of an ancient collapse scar known as English's Crater, but it was unsupported to the east, where the terrain simply fell away to the sea down the Tar River valley. Red-hot rock continuously sloughed off the ever-expanding steep face of the dome, feeding a hot, unstable scree. Every few weeks, whole segments of the accumulating mass disintegrated, generating *nuées ardentes*. All of us were aware that it was the explosion and destruction of a lava dome on Pelée that had sealed the fate of St Pierre in 1902. The situation was very volatile, and I sensed a supercharged mood of excitement and trepidation.

I was staying at a villa in Old Towne rented by the observatory. There was a veranda with wicker chairs overlooking an exuberant garden and a swimming pool – quite an upgrade from my college apartment in Cambridge, where I had recently been hired as a lecturer. The lawn was well tended, the flame trees in bloom – only the ash in the pool hinted at abandonment. Most mornings, my breakfast on the terrace was interrupted by a call of, 'Buy my juicy mangoes!' from Beryl Grant, known to all on the island, and then in her early seventies. She didn't bother to conceal the act of harvesting the fruit on the property, but it was impossible to refuse her sales pitch. Like many other farmers on Montserrat, Beryl had been relocated from the east side of the island, where much of the produce was grown. So many lives and livelihoods had been turned upside down, and no end to the crisis was in sight, yet most remained resolute, sustained by deep Christian faith.

My role at the observatory was to support the surveillance of gas emissions from the lava dome.[21] Working alongside my new colleagues, I used an ultraviolet spectrometer called a Cospec to measure the output of sulphur dioxide from the volcano, and an infrared spectrometer to determine the proportions of several different gases.[22] The Cospec was connected to a chart recorder, and a telescope, which I directed up at the sky out of a car window. I would drive back and forth beneath the plume drifting from the lava dome while keeping an eye on the pen registering the sulphur dioxide overhead by its absorption of ultraviolet light. After a few traverses, I would head for St

George's Hill, overlooking Plymouth, and stand on tiptoes to hold aloft a simple anemometer to measure windspeed, which factored into the calculations. Ashfalls sometimes hindered the work, but I aimed to make daily surveys, which took a couple of hours. The real labour, though, was the subsequent number crunching, which involved counting thousands of tiny graph paper squares beneath the ink trace of each run. At last, I would have an estimate of how many tonnes of sulphur dioxide were coming out of the volcano each day. (It was the toil of using the Cospec that motivated me to trial a new generation of compact digital spectrometers a few years later at Masaya volcano in Nicaragua.)

Operation of the infrared spectrometer was even more demanding and required two people. It had to be aligned precisely with the light from an infrared lamp positioned some distance away – this could take a frustrating hour of 'left a bit, up a bit' instruction over walkie-talkies. We needed volcanic gas to drift between the lamp and spectrometer, and so sought places where some of the airborne plume was grounding leeward of the summit. This took us to Plymouth, by now a ghost town with light ashfall drifting up empty streets taken over by dogs. Eerier still, we operated the equipment in deserted rural areas inland from Plymouth. One location where we were suitably fumigated was Upper Amersham, near an old sugar mill tower. The abandoned farms looked inexpressibly godforsaken. All vegetation had been poisoned and metal roofs eaten away by the gases we were trying to measure. We encountered a depressed pig that had turned yellow, presumably from the acidified rain falling through the gas plume. But the fact it was tethered and still alive meant that someone was coming regularly into the exclusion zone.

I felt a bit envious of the seismologists, whose sensors, once installed in the ground, continuously transmitted readings by radio link to the volcano observatory. But it was worth the effort to obtain the gas measurements as they were yardsticks for the depth and volume of magma still below ground. I would present my latest findings at tense staff meetings, which took place in the observatory three times a week. Many signals were picking up dramatically, including earthquake activity, expansion of the lava dome and sulphur dioxide emissions. It was obvious that the volcano was becoming more active, but what did that mean for how things might play out over the next days and weeks? Was it prudent to expand the exclusion zone around Soufrière Hills? The discussion was often heated and I have never felt greater responsibility or urgency to figure out the meaning of my data.

I was aware, too, of situations where scientists had clashed over the evaluation of volcanic hazard – famously during a crisis in 1976 on the island of Guadeloupe, situated between Martinique and Montserrat. Another outpost of France's *Outre Mer* territories, Guadeloupe's La Soufrière volcano had begun simmering in July that year, with escalating seismicity, and explosions that plunged the capital into darkness as ash clouds drifted over.[23] Haroun Tazieff (whom we left on Stromboli with the dying Muratori in his arms) was there, leading a team of scientists recording gas emissions and earthquakes. While more given to popularisation than Perret, he shared the same deeply empirical and experiential approach to reading volcanoes' moods. Based on what he'd observed, Tazieff announced that, though the situation demanded ongoing surveillance, a major eruption was not imminent. But another group, led by a specialist on the (inactive) volcanoes of the Auvergne, and backed by Claude Allègre, the new young director of Tazieff's institute in Paris, feared a 'monstrous' and 'catastrophic' eruption on the scale of Pelée's in 1902.[24] When a hint of fresh magma was identified in the ashfalls, the precautionary view prevailed, and the prefect of the island ordered the evacuation of 70,000 residents – a quarter of the island's population.[25] A vitriolic duel ensued over the prolongation of the evacuation.

The acrimony between Tazieff and Allègre filled columns of French print media.[26] Sadly, it pitted experience, intuition and empiricism against models, probability and uncertainty, when these different perspectives and paradigms might have been integrated and harnessed to the benefit of the threatened community. After several months, La Soufrière went back to sleep, and the refugees were sent home. For many, Tazieff was vindicated. He once wrote that volcanologists should keep 'as cool as a cucumber' in a crisis.[27] But Allègre surely pickled Tazieff over the Guadeloupe stand-off, removing him from the post of head of volcanology at the institute.

It might seem sensible always to err on the side of caution as an operational plan, but displacing a population from their homes and livelihoods – even at the smallest scale of a village like Talabre in Chile, when it was threatened by Láscar volcano – is immensely disruptive. The cost of the Guadeloupe crisis was estimated at $500 million USD, and left the economy in tatters with much of the population reduced to 'utter want'.[28] Eruption forecasts are highly uncertain, sometimes contested. Managing the menace – and ultimately, perhaps, facing the dilemma of whether or not to evacuate thousands of people – demands clear-headed understanding of scientific uncertainties on the part of the authorities, and an effective and lasting co-operation of state and society.

The case of 'La Soufrière '76', with its duelling scientists, is considered a masterclass in how to mismanage a volcanic crisis.[29] On Montserrat decades later, I was shocked, then, to discover rifts and enmities between scientists at the observa-tory.[30] The memory of Pelée had hung over Guadeloupe and the memory of both overshadowed Montserrat. At the same time, the confrontations revealed another facet of the challenges of dealing with a volcanic emergency: personality clashes, and differences in background, management style or risk tolerance can make it very difficult to reach consensus. Who are the authorities and public to believe when the scientists disagree and engage in character assassination? Whose expertise counts most?

While interpersonal friction fuelled tensions in the observatory, the coexisting peril and excitement of the crisis surely heightened everyone's emotional state. One of the distinctions of volcano observers is that they usually belong to the communities they serve to protect. During emergencies, they may well fear for their own safety when they enter exclusion zones to deploy instruments or collect rocks. For the Montserratian staff, their own families were, in a sense, under their safeguarding. These personal, potentially existential circumstances bring a unique frisson to the work.

Some of the discussion at observatory meetings focused on hazard zonation on the island. This determined where people could live or work, and where was altogether off limits. After a year of crisis, several thousand Montserratians had settled elsewhere in the Caribbean, in Canada or in the UK. Only 7,500 people remained on the island. With such a prolonged affair, some argued it was time to make finer-tuned decisions concerning access to allow the lingering population to 'live with the volcano'. There were also practical and political considerations that somehow trumped objective safety concerns. For instance, there was no appetite to close the airport – a lifeline to the outside world – despite its proximity to the Tar River valley. The outcome was a 'microzonation' of territory and a complex map delineating gradations in estimated risk. This was used in conjunction with a system of six alert levels. With frequent updates reflecting evolution of the volcano's behaviour, this scheme became increasingly challenging for anyone to understand.

The observatory was already engaging a wide range of surveillance techniques. These included GPS surveys and laser ranging to track subtle expansions and contractions of the volcano, and use of seismometers sensitive to different vibration frequency bands. Paradoxically, while miniscule movements of the crust could be detected, and earthquakes located with precision deep underground, the most mundane but vital requirement – visual observation of the dome – was near impossible. This was because the summit was sheathed in cloud much of the day. Fortunately, the senior scientists had occasional access to military satellite radar images that revealed prominences and scars on the dome, and the extent of new deposits in the Tar River valley. But it remained frustrating not to know hour-by-hour which parts of the dome were propagating. Later, one of the observatory staff hit upon the simple but brilliant idea of setting up a camera connected to the observatory by a radio link. It took a photograph every minute. While no observer could sit and watch for twenty-four hours, waiting for a brief clear view, the automated system would capture those rare moments.

The populous west and windswept east sides of the island seemed like different worlds. From Old Towne to Cork Hill in the west, the road was fringed by tropical palms, breadfruit trees, hedgerows of bougainvillea and hibiscus. Mango trees dripped overripe fruit on to the road. But approaching Harris, across the corrugated backbone of the island, and on to Bethel, there were vegetable gardens and pasture grazed to millimetre length by scrawny brown and white goats. What few shrubs and trees grew here were stunted and bowed by ceaseless wind. The sea, too, was leaden, merging into haze. Lately, the *nuées ardentes* had been crossing the shoreline, skimming with remarkable ease over the waves and leaving behind a steaming delta of lava boulders. Plymouth occupied some of the only level ground on the island, hence its situation, but this very flatness arose from the same processes of debris accumulation during past episodes of volcanism. With hindsight, the mere fact it looked ripe for settlement was a geological warning sign.

Working on the dome itself was out of the question, but I persuaded the observatory chief to allow me some helicopter time to operate a spectrometer from the air to capture stronger gas readings. We made several passes close enough to feel fierce radiance through the open window. The steep face of the dome was a precarious jumble of car-sized angular and scaly blocks of grey lava. Dust trickled off it continuously, and a rocky spire protruded from the top of the unstable cupola. These overactive pinnacles formed within a day from focused extrusion of the ultra-viscous magma, but usually collapsed another day or two later. I was acutely aware that I needed to be quick. The pilot advised me there was no safety margin to deal with an engine failure – we'd plummet into the hot pyroclastic deposits. In hindsight, I chalk up this experiment alongside my first measurements at the edge of Stromboli's crater – not worth the risk. The results showed some promise but ultimately made no contribution to the monitoring efforts.[31]

Soon after, the eruption entered a yet more ominous phase. A new spine had been seen, rising as high as the Tower of Pisa, and chunks of lava as big as a bus had travelled a long way down the valley. *Nuées ardentes* were reaching the sea with greater ease, since much of the tree cover had been stripped off the lower flanks of the volcano. All this commotion was accompanied by menacing ash clouds, which didn't go unnoticed by the remaining residents. People sensed change. I encountered an eighty-one-year-old woman in Plymouth, who told me: 'I never seen this before – it is the world coming to an end. If it can happen here on Montserrat, it can happen in another place – we must trust in Jesus.' The island governor, Frank Savage, was not one to cede oversight of the crisis, however. He was alarmed by the latest growth spurt of the dome, which was now higher than the old crater wall, beyond which there were still many people farming on the north flank of the volcano in an area called Farrel's. Would there be time to evacuate them if the dome collapsed in that direction?

Savage also felt the observatory's hazard bulletins were being misunderstood or ignored. 'They're too bland,' he argued. 'They need to be written in layman's terms giving the gist of the situation. If people don't listen, we're wasting our time.' One of the Montserratian staff, Pops, countered that many on the island no longer wanted to hear what the scientists were saying anyway – they just wanted to stay in their homes.[32] As many as 500 people were then living in hazardous areas. They reasoned that their farms were still there, and government agencies were still buying their produce to distribute in shelters for evacuees. Pops knew of a farmer who insisted he would only move when he saw the rockfalls coming at him. People were monitoring the volcano for themselves, he continued, adding that 'to get people out we will have to enforce it'.[33]

I encountered this mindset in the Tar River Estate, on the north-east flank of the volcano and overlooking the valley. I had joined a TV documentary crew from Canada who were filming right at the margin, which had been scorched by *nuées ardentes* the previous day.[34] In the valley below, shrubs were raked over in the direction of the flows, and trees blasted flat, only their stumps remaining. The dome felt unnervingly close – I have never had a stronger sensation of looking down the barrel of a gun. *Nuées* could have reached us in thirty seconds. Astonishingly, we found a farmer there. His name was Prince, and he had been relocated to the north of the island, where he was working on construction of temporary shelters. 'I lost five cows yesterday,' he lamented, as he led us to their bloated, charred and ash-coated corpses. 'They usually know if the volcano is going to blow, and run. I can't complain to no one. I just have to survive.'

Compounding the problem on Montserrat was the protraction of the crisis and stop-start behaviour of the volcano. One of the chief scientists who worked on Montserrat observed, 'the island is exactly the wrong size for an eruption'. Any smaller, and it would have been the obvious choice to evacuate the entire population; any larger, and there would have been better options to resettle people away from the volcanic threat and rebuild the economy. Instead, an unstable compromise between man and magma has prevailed on Montserrat more or less ever since the eruption began.

Around noon one day, I had finished lunch at the villa and was about to walk back to the observatory when I heard chatter on the walkie-talkie about an eruption cloud – I raced to the veranda and saw a towering black plume almost directly overhead. Small accretions of ash were already falling out. Then I spotted a diffuse white mist expand very rapidly outwards and downwards. I didn't know what to make of it, and it momentarily terrified me. But I recalled Perret's description of 'flashing arcs' that he had seen accompanying explosions, and realised it was a shockwave. Heavy ashfall ensued, accompanied by thunder and lightning, and I raced around the house, closing doors and windows. The power cut out and it was instantly pitch black – I could make out nothing. I stood in the dark and waited. When the cloud dispersed, I looked at the garden: so lush minutes before, it was now greyed out by a film of fine ash. The air smelled like bleach and damp burned stubble.

Seen from the airport on the other side of Montserrat, the cloud had towered over the island and several *nuées ardentes* had reached the sea. This was the biggest event yet of the entire episode. So much rock had crumbled from the dome, the crater looked like a grin with a missing tooth. Huge boulders had reached the tongue of debris stretching into the sea. In upper parts of the Tar River valley, the erosive force of the flows had planed down ridges and hills, whereas downslope they had infilled gullies and channels with lava blocks and ash. For a while, it felt like the end of the world, but as the dust settled it emerged that no one had been hurt. In a matter of days, the giant wound in the dome had healed with new simmering lava.

*

A year on, and a full two years since it began, the eruption was still showing no signs of easing up. The volcanologist Thomas Jaggar once defined geology as 'a sense of slow motion', but here on Montserrat the pace of geological change was frenetic. Plymouth looked like it had been set in concrete. Only the upper storeys of buildings stood proud of an alluvion of boulders and grit; you could only guess where streets had been.

In May 1997, a team from the observatory made a daring mission to deploy a device called a tiltmeter on the old summit of Soufrière Hills. Such sensors are now everywhere – for instance in smartphones, to tell what orientation they are being held in. Highly sensitive tiltmeters can measure changes in inclination equivalent to raising a one-mile-long bar by less than two millimetres at one end. The tiltmeter the team installed was connected by radio link to the observatory and gave a continuous measure of the subtle swelling and sagging of the volcano. This revealed a remarkably regular oscillation every six to nine hours: the volcano had a pulse.[35] Nothing like this had been observed before on such an explosive volcano. I couldn't help thinking of Perret, who was always seeking design in data – he would have been mesmerised by these observations! The rhythm became so routine that the observatory could plan fieldwork around it and make reliable short-term forecasts of explosions and *nuées ardentes*, which followed dome inflation.

The pattern continued until shortly before 1pm on 25 June 1997. Another dome collapse was underway, but this time it excavated deep enough to trigger a runaway process that exposed highly pressurised, gas-rich lava, which made for far more mobile *nuées*. Additionally, the dome at this time had already begun spilling over the old ramparts of English's Crater – the debris shed from it was no longer confined to the Tar River valley. Over the next ten minutes, three successive *nuées* pulsed down the north and north-east slopes of Soufrière Hills at up to fifty miles per hour. They overran farmland and villages which I had often passed through, razing buildings to their foundations and igniting fires that blazed for hours. Eighty people were in the exclusion zone at the time; nineteen of them perished. Among them was Beryl Grant.[36] I wasn't on the island at the time, but the news came as a great shock.

Of the survivors who were interviewed, almost all commented on the acute silence of the lethal flows. This account from one eyewitness captures the experience:

> Ash . . . came down sly and sneaky. It travelled faster than a car . . . The pieces of fire were small and it was there before the place turned black. Was grey and black, one mighty cloud . . . No hear a thing before I saw it . . . Start to feel hot, hot, hot, like I was in an oven. The whole place turn black for about 20 minutes, heavy ash falling . . . black like black cloth . . . Smelt strong, like lighting matches smell . . . Fire was pitching on the pavement by the school. Fire was pitching in the air as well. The whole of my body felt hot. There was a heavy rushing wind with it . . . I think it must be the end of the world.[37]

While the *nuées ardentes* may have been mute at the surface, the ground reverberated to the percussion of fracturing and colliding rock and the chorus of magma rising to fill a new chasm in the dome. A colleague made seismic signals recorded on Montserrat audible by shifting their frequencies. It makes for an extraordinary symphony of sirens, whipcracks and drumrolls. Perhaps this is what Perret (who, more than anyone, *listened* to volcanoes) could hear when he connected his earphones to the earth-contact microphone.

The case histories of both Pelée and Soufrière Hills provide tragic illustrations of how science is not enough to protect communities menaced by volcanoes. If it is to benefit society, it has to be translated into effective decisions; authorities must be persuaded to listen, absorb, react. Perret's brilliant monograph on the Soufrière Hills' earlier crisis was forgotten, and the prescient 1980s hazard report ignored.[38] In the case of Pelée, a 'scientific commission' of sorts had been set up before the 1902 calamity, but despite unease on account of the volcano's mounting agitation, the committee's public statement was bland enough for an op-ed in a popular newspaper to downplay the threat. 'Where could one be safer than in St Pierre?' it quipped, actually prompting hundreds of anxious countryfolk to seek refuge in the doomed city.[39] They, along with the paper's press, its proprietor, its journalists, and 27,000 others, were wiped out the next day.

While opportunities for mitigating disaster were missed, the scientific attention subsequently directed at the two Caribbean volcanoes turbo-charged volcanology as a discipline, stimulating developments both technological and humanistic. Following Pelée's eruptions, *nuées ardentes* were scientifically recognised and described for the first time. On both Pelée and Soufrière Hills, Perret pioneered 'multiparameter monitoring', the combination of geophysical and geochemical tools for volcano surveillance. This is standard practice today: the most reliable eruption forecasts draw on simultaneous observations of earthquakes, topography and gas emissions.

Perret also believed that to understand volcanoes required 'the wholly indescribable *feeling* which experience alone can give'. Although this speaks to the truth that scientists are not detached observers and theorists, but rather people with instincts passionately trying to understand the world, it is also the point where dangerous ground looms – *feelings* humanise us, but they can also remould opinions as convictions, supplant impartiality with infallibility. Haroun Tazieff, whose experience can justly be compared with Perret's (but whose self-belief far outstripped the American's), felt vindicated in his call to end the evacuation on Guadeloupe in 1976. Some (with hindsight) agreed, yet we can also sympathise with Claude Allègre's dilemma, when he said 'I don't want to play Russian roulette with the lives of 70,000 people.'[40]

There have been many situations in which biased or sloppy judgements of risk or ulterior motives swayed decision-making. To counter these problems and avoid turf wars between scientists, a method called 'expert elicitation' has been used to harness subjectivity and uncertainty when enumerating volcanic risk. In fact, its application was pioneered on Montserrat by a geophysicist, Willy Aspinall, who had witnessed first-hand the venomous clash of scientific opinion on Guadeloupe in 1976.[41] The method involves tasking a panel of specialists to draw on their experience and intuition to estimate probabilities for different scenarios of how a volcanic crisis might unfold (from relaxation to escalation). Statistical techniques are then applied to combine all the viewpoints. The skill (and obstinacy) of individual experts is gauged in advance through a test (for which answers are known), so that their influence on the calculations of risk can be weighted appropriately. Colleagues who took part told me: 'It isn't quite what I would have come up with, but I can sign up to it.' Such collective responsibility takes some weight off a chief scientist's shoulders, too. That must be a good thing – even with the degree of answerability I felt as an intern at the Montserrat observatory, I cannot imagine the stress of the job at the top.[42] Commenting on the moral quandary *he* experienced during the 1930s crisis, Frank Perret confided that 'such anxiety and responsibility . . . cannot be imagined by those who do not know'.[43]

Perret was an optimist, though. Late in his career, having established a Centre for Volcano Study in St. Pierre, he wrote that: 'an enormous amount of valuable research has been, and is being, effected . . . giving the directly essential power of *diagnosis* and *prediction* [of eruptions]'.[44] While the vagaries of volcanoes will always defy easy pronouncements concerning their future conduct, progress continues.[45] What gave Perret deep satisfaction in his vocation is still felt by volcanologists today – the worldly desire to answer 'humanitarian needs', but also the personal revelation of our beating Earth.

Night Market of the Ghosts

Propitiating the spirit of the volcano.[1]

'Their story is buried in the forests that lie silent as the sleeping crater's mouth.'
  – Franz Wilhelm Junghuhn, *Java, seine Gestalt*[2]

Being closer to the heavens, capturing life-giving water from the clouds, mountains have long been considered sacred places. Volcanoes, whose singular peaks soar above fertile plains and valleys to pierce the clouds, are also abodes of ancestors and souls, and sites of miracles – but more than that, they are portals to infernal realms of gods and spirits whose supernatural interventions in human affairs manifest in ash, fume and flame. Where nightly fire glow tinges the starry sky and thundering gases reverberate the air, it is no wonder igneous nature is engrained in indigenous knowledge.

Nowhere is this more true than in Indonesia, which straddles a section of the Pacific plate's subduction zone, and is home to more active volcanoes than any other nation. Here, wild nature sweeps skyward from dense, entangled urban and rural fabrics.[3] In Indonesian, they are *gunung api*, 'fire mountains'. The anthropologist Peter Boomgaard explained their significance simply: 'We can, without exaggeration, regard the mountains of Java as the basis for its society and its culture . . . Indonesia as a whole is shaped by its mountains, its volcanoes.'[4] More than seventy have blazed in past centuries; each home to ghosts and spirits, venerated and feared in equal measure. I came to know several of them – my first encounters with volcanoes – during gap year travels between high school and university. For me they inspired wonder, not dread.

That curiosity was elevated by reading the one book I had brought with me from home: *Volcanoes*, by Peter Francis. I did not guess then that five years later Peter would be my PhD advisor, but his writing brought another layer of meaning to the sheer spectacle of volcanoes. Crossing the Indonesian archipelago, I saw nature and culture collide in landscape, belief and custom, sensitising me to the charisma and mystique of volcanoes. But it was thanks to Peter's book that I also began connecting culture, economy and drama with rocks, resources and topography.

Near the end of my trip, I crossed Flores in the country's south-eastern islands, which look like roughly-fitted jigsaw pieces on a map. They are laced with volcanoes built up from the sea. I had been moving east for months, and now, riding a bus to the coastal town of Larantuka, I neared my trail's apogee. The road was in terrible condition and the driver had to repair several punctures. Halfway, we approached an imposing double-coned volcano, Lewotobi, its pinnacles named *laki laki* (men) and *perempuan* (woman). At the foot of the volcano, a nun got off the bus to visit a seminary partly concealed behind coffee bushes and shade trees. I noticed their leaves were filmed with a pale powder, but thought nothing of it at first.[5]

I took a ferry to Weiwerang on Adonara island the next day. The vegetation was dustier still, and a market stall owner explained that the nearby volcano, Ili Boleng, had recently exploded, with clouds of ash and fiery rock. At once I understood what I had been observing was not the result of passing vehicles stirring up road dust. I resolved to scale the mountain in the morning. A family – seemingly with underworld connections, since their surname was Demon – put me up in the village of Lamalouk.[6] A nourishing breakfast of green pea porridge in my belly, I aimed for the volcano, whose summit, sometimes poking through cloud, loomed beyond a nearby stand of tall hardwoods. In the next village, a woman implored me to turn back: 'There is danger up there,' she said. But I had tackled higher volcanoes on Java, I told myself, and could handle this.

The path faded quickly, ending in thickets of lacerating *alang-alang* grass coated in dew-soaked ash. My legs and arms soon stung from wounds inflicted by the weed's barbed blades, and I diverted towards a gully. I imagined vast quantities of muddy ash sweeping down it in the rainy season, imparting nutrients to soils. This was easy terrain to follow at first – a layer of soft ash on bedrock – but as the slope increased, the ash concealed not solid ground but mobile scree. I could only advance by scurrying on all fours slightly faster than I was sliding back. The gully became more of a canyon, with unscalable sidewalls revealing generations of ash layers, and I was now dislodging disturbingly large masses of rubble. Damp clouds shrouded the slopes, and I felt the cold, dressed in shorts and a T-shirt. It wasn't the best-planned expedition.

Retracing my steps was the only option, but not far below an uprooted tree that had tumbled down was leaning close enough to the top of the channel for me to escape. Reaching the end of the bole, I found a ghostly, petrified scene – thick ash underfoot and a thin forest of spindly tree stems stripped of foliage, branches, bark and all. I continued up, but again ran into steeper slopes of ash and gravel on which I could get no purchase. I persisted on hands and knees and, after another hour of toil, reached the crater rim.

Wind gusts flung ash in my eyes, lustreless clouds swirled, and more dust fell from grey vapours lofting from the chasm as I skirted the rim. Only fleetingly, I wondered about the village woman's premonition. I had a stronger feeling, though, of reaching a turning point. I had been journeying east for months, and now it was time to backtrack, to return home for the start of the university year. But the significance went beyond a geographical pivot: to have reached the summit of a far-flung erupting volcano, to see my footprints in pristine cinereous powder that a day before had been viscous magma in the Earth's interior, and to sense my own transition from childhood to adulthood, brought a self-awareness that was new to me. A nomadism and curiosity that still influence my career were ignited beside this expressive, mysterious crater. Volcanoes were now engrained in my story.

★

Indonesia was also the right place for another aspiring scientist, Franz Junghuhn, to be switched on to volcanoes. Born in 1809, the same year as Charles Darwin, Junghuhn's early challenge was to sidestep the medical career mapped out for him by his father, for his true passions were botany and mycology.[7] He was so precocious that by age twenty, he had published a creditable work on fungi of the Harz mountains (written in Latin) in the era's leading botanical journal.[8]

After many travails, Junghuhn enrolled at the Royal University in Berlin in 1830. He would have been thrilled at the prospect of attending lectures by Alexander von Humboldt and getting his hands on the great naturalist's collection in the herbarium. But it was not to be; soon after matriculating, Junghuhn found himself duelling with pistols after an altercation with another student in a beer garden. Junghuhn took a bullet in the leg while missing his opponent, who subsequently killed himself, so great was his sense of dishonour. That might have been the end of the matter, but news of the shoot-out reached the authorities. Duelling was rigidly prohibited, and Junghuhn was handed down a ten-year sentence to be served at Ehrenbreitstein Fortress, built following Napoleon's defeat (and just a stone's throw from Laacher See, the site of a huge volcanic eruption 13,000 years ago).[9] Twenty months into his sentence, his pleas for clemency unanswered, and now ensnared by a more alarming botanical fixation – opium – a desperate Junghuhn could endure no more. He fled one night through a window in the laundry room, likely with inside help. Ironically, just days later, Junghuhn's captors received a royal decree stating that their erstwhile inmate had served sufficient time and should be released. But by this time, the fugitive had reached France.

Following a stint as a sergeant with the French Foreign Legion in Algeria, which opened his eyes to the evils of colonialism, Junghuhn returned to Europe. Back in Prussia, he was reunited with the one man, a teacher and botanist, who had befriended him during his confinement at the fortress (and who had probably procured the key to the laundry room). The pair made an excursion to Laacher See, whose 'wild melancholy' and 'traces of chaotic devastation' aroused Junghuhn's romantic spirit. Perhaps this *Feuerberge* (fire mountain) sowed a seed, just as my climb of Ili Boleng did.[10]

A year later, in 1835, Junghuhn arrived in Indonesia, with a commission as a medical officer. His spirit soared as he sighted Java's tropical vegetation from the ship. But soon he would become equally passionate about the volcanoes. He mapped them, took rock samples, measured temperatures and altitudes, and evaluated soil types and vegetation zones. He also documented the Javanese veneration of volcano gods. Indeed, as one historian has said, 'Looking into a volcano was, for Junghuhn, essentially looking God right in the face.'[11] He recognised the duality of volcanoes in cycles of destruction and regeneration – the prolific biodiversity and agricultural fertility conferred by volcanic soils and ashfall, but also the terror and split-second devastation of a pyroclastic eruption. Finding, at the feet of volcanoes, ruined pantheons with statues of the gods to whom they were dedicated lying shattered and forsaken, tangled in figs, Junghuhn understood, too, how the lives and shifting cosmologies of the Javanese are bound up in transformative nature.[12] Volcanoes, he was told, were guarded by tigers and snakes – real threats to be sure, but also, I suspect, metaphors for stealthy *nuées ardentes* and serpentine smoke trails rising from fumaroles.

Vision was Junghuhn's critical sense; his eyes were bright and sharp like a raptor's.[13] Accentuated yet further with a trusted telescope, his acute eyesight helps to explain the quality of his surveying and cartography. His nature writing, too, is both visual and expressive, described by the author Eric Beekman as alternating 'between seduction and instruction . . . closer to an art than a technique'.[14] Junghuhn's published syntheses of the geology, flora and fauna, and culture of Java are encyclopaedic landmarks, and his belief in the unity of nature is inspiring. Moreover, he was sympathetic to local knowledge, recognising that historical events are veiled in stories attached to landmarks.[15] Above all, I admire his sketches – some transformed into enthralling lithographs for his 'Java-Album'.[16] Their aesthetic and scientific qualities recall the illustrations in Sir William Hamilton's *Campi Phlegræi*, but go beyond the naturalistic precision of the Italian scenes, employing subtle artifice to elevate their fidelity, through a bird's-eye view, for instance.[17]

Merapi, in central Java, was the first active volcano both Junghuhn and I climbed. It is among the most dangerous and significant in Indonesia, with over a million people living in its shadow. The country's last Sultanate, a relic of the former kingdom of Mataram, radiates from the volcano's peak. The sultan's palace, the *kraton*, lies on an axis connecting the spirit realms of sea and volcano with the sublunary world, linking the eternal and ephemeral. It is only twenty miles south of the summit, in the heart of Yogyakarta, a city of half a million people. In an echo of ancient temple architecture, the *kraton* is a miniature replica of the cosmos. Merapi's eruptions are often deadly owing to the high population density on the volcano's flanks. Two very different approaches – spiritual and scientific – seek to lessen the menace.

For many Javanese, Merapi is a kingdom of ghosts. Afterlife is a mirror of the material world, but it exists below ground – a rice farmer here will be a rice farmer there. The underworld spirits, watched by their overlords Empu Romo and Empu Parmadi, can warn the living of an impending eruption. An earthly gatekeeper, the *juru kunci*, is appointed by the sultan to read the omens. He lives in the village of Kinahrejo, dignified by its purpose as portal, but perilously close to the crater. His daily observances are meant to sustain an equilibrium in nature. Once a year, a more elaborate ceremony takes place, involving dozens of the *kraton's* courtiers, with offerings of flowers, rice, incense and money made beside a sacred rock in the forest above Kinahrejo. If these performances aren't done correctly, the gods will be displeased and an eruption might follow.

From 1980, Merapi's mediator, then fifty years of age, was Mbah Maridjan. He had followed his father's vocation and conducted his formal affairs with the invisible world with solemnity. His popularity soared after he refused to evacuate during a lethal eruption of Merapi in 2006, despite appeals from the national volcanological agency, the Sultan of Yogyakarta and even the vice president of the country. He lived to tell the tale, which was relayed with relish by the media. Mysticism had trumped modernity (though that did not prevent Maridjan from accepting a commission to appear in commercials for an energy drink).[18]

Merapi soon regained its composure, deflating after its evacuation. But just a few years later, in mid-September 2010, seismometers on the mountain recorded an uptick in tremors, precise surveys revealed a bulging summit, and geochemical sensors registered escalating gas emissions. Magma was on the

ascent, and the volcano alert level was raised from 'normal' to
'on guard'. At this time, I was involved in research on Merapi
with French and Indonesian scientists. I provided one of our
bespoke ultraviolet spectrometers from Cambridge to measure
sulphur dioxide emissions from the crater, and my colleague,
Marie Boichu, trained local observatory staff in its operation.
My Indonesian collaborator, Sri Sumarti, who was responsible
for gas-monitoring at Merapi, told me later that the equipment
played an important role in assessing the volcanic threat. With
both increasing gas fluxes and shaking of the earth, the alarm
was raised further on 20 October (to 'be prepared'), and to the
highest level on 25 October, signifying an imminent eruption.
During all this build-up, Sri and her team at the observatory
stepped up their engagement with the communities on the
mountain to explain the signals and their implications. 'We call
this socialisation of volcanology,' Sri explained.[19]

Tens of thousands of people were evacuated, but once more
Mbah Maridjan refused to leave his post. This was his last stand;
on the following day, blistering *nuées ardentes* swept through
the village with hurricane force. His body was found in the atti-
tude of prayer, fixed by the intense heat.[20] The destruction
of the village of Kinahrejo was near total, yet so transient was
the passage of the *nueé* that Mbah Maridjan's batik clothing was
unburned. A veneer of fine ash covered the scene. An even
larger eruption followed just over a week later but by then
more than 400,000 people had sought refuge beyond the danger
zone.

I visited Kinahrejo two years later and found a booming dark
tourism scene, with off-road jeep trips to the devastation zone
and other attractions.[21] In another twist of modernity, Mbah
Maridjan's widow, who had not been in Kinahrejo on the fateful
day, was sitting in a kiosk selling instant noodles and T-shirts
depicting her late husband.

I spoke with the new gatekeeper, Mbah Maridjan's son, Pak Asih. Standing beside his incinerated motorbike, displayed for the tourists, and close to his father's tomb, he explained it remained his duty to communicate information on the volcano from the spirit world. However, he saw this as complementing, not negating, scientific data analysis and knowledge. He seemed to accept that during the 2010 crisis, observation had bettered observance. The evacuations had saved tens of thousands of lives. In fact, there were even calls for the director of the national volcanological agency, Pak Surono, to be made the new *juru kunci*, a standout moment in the culture clashes between science and spiritism.

But years later, Surono was close to tears as he told me of the profound responsibility he still felt for the losses: 'Despite our differences, I cried when they told me Mbah Maridjan was killed. Science is for the head, but the heart also has to understand the sentiments of the people. Volcanology is about the people.'[22]

★

The underworld spirits are not universally placated; the signs of disquiet may go unheeded. So it was in 1815 on Flores island's neighbour to the west, Sumbawa. But this was an eruption up to 1,000 times larger than that of Merapi in 2010. Junghuhn described it as 'one of the most terrible ever to have occurred on the whole earth according to human memory'. Two centuries on, this epithet holds. Its grim effects were felt not only across the Indonesian archipelago but also in Europe and North America, where it impacted millions of lives by giving rise to 'the year without a summer' of 1816. The unseasonable chill and its consequences kindled fresh folklore across the world.

The protagonist, Tambora, forms a peninsula on the north coast of the island, and may have been the highest peak of the whole archipelago before the cataclysm. Indonesia had become embroiled in the Napoleonic wars, following the defeat of its colonial masters, the Dutch, by France. This compelled the British, whose imperial possessions included the Malay peninsula, to launch a brief and successful naval assault. So, when Tambora exploded, the lieutenant governor of Java was Sir Thomas Stamford Raffles, a luminary of the British Empire, and as enthusiastic about butterflies and other natural wonders as he was adept at military engagement and colonial administration. Thanks to his attention and interventions, along with other testimonies, we can piece together the events and their consequences in the region.[23]

At the end of the 1990s, I was researching a book, *Eruptions that Shook the World*. For a change, I put the spectrometers away, eased back (a little) on fieldwork and spent days in the university library. Had any of my friends or colleagues been looking for me then, they would never have suspected I was in the rare books room of the library. But that was where the trail to any and all intelligence on Tambora had led me.

Among the accounts, I found a fascinating vignette attributed to Raffle's colleague John Crawfurd, resident governor of Surabaya province in north-east Java. In 1814, he had sailed past Tambora and noted dark, threatening clouds coming from its direction. He misread them for an approaching squall until ash began settling on the deck.[24] With hindsight, this activity signifies the reawakening of the volcano from prolonged slumber. Magma that had accumulated in a deep reservoir had been pressurising due to bubble formation and was now rising, unlocking sealed conduits, seeking escape.[25] Perhaps the communities on the volcano's flanks were moved to make offerings to the restless mountain.

The action escalated dramatically on the evening of 5 April 1815, with a violent pumice eruption whose concussions reverberated across Java, where they were universally mistaken for hostile fire. In Surabaya, it was thought a merchant ship was under attack from pirates, and gunboats were launched to intervene. Only the next day, when ash settled across the land, was the mystery solved. Things were much worse around the volcano, with crops flattened beneath ash. The colonial Resident of Bima, fifty miles from Tambora, dispatched a deputy named Israël to gauge the situation. But the recent explosions had tapped just a fraction of Tambora's compressed magma – the real paroxysm came five days later.[26] Crawfurd, two islands away, was forced to conduct his daily affairs by candlelight as a gigantic ash cloud plunged the whole region into Cimmerian gloom. Poor Israël was never heard from again.

The scale of the catastrophe can be partly gauged by how little information reached Batavia (Jakarta) afterwards. Eventually, Raffles sent an envoy, Owen Philipps, to Sumbawa to distribute foodstuffs and ascertain the facts of the eruption and plight of the people. The emissary met the head of one of the villages closest to Tambora, whose survival is remarkable as it's likely almost all on the peninsula perished. He told Philipps that at around 7pm on 10 April, three columns of flame burst from the mountain, uniting above it at great height 'in a troubled and confused manner'. Then:

In a short time, the whole mountain . . . appeared like a body of liquid fire, extending itself in every direction . . . and soon after a violent whirlwind . . . blew down nearly every house in the village . . . tearing up by the roots the largest trees and carrying them into the air, together with men, horses, cattle.

In light of the scientific studies following the tragic 1902 eruption of Montagne Pelée, it is certain this is a description of *nuées
ardentes*. Now the magma reservoir was uncorked and the eruption on track to its once-in-a-millennium ranking. So much
molten rock was disgorged that the mountain toppled in on itself
above the voiding chamber, leaving an immense yawning hole in
the ground. An apocalyptic fallout of ash across Indonesia's
south-eastern islands stifled crops and razed dwellings.

Franz Junghuhn was only five when Tambora burst open, but
the disaster remained fresh in the collective memory of the East
Indies when he landed in Java twenty years later, and he was
among the first to synthesise the available information. As many
as 12,000 people were annihilated instantly by the *nuées ardentes*,
but over the following months many tens of thousands more
perished – as far away as east Java – from starvation and disease.
Before the eruption, Sumbawa had been famed for its lush
valleys, coffee and honey, fine horses and timber.[27] After, it was
a scene of unimaginable misery, roadsides littered with corpses
and shallow graves, villages deserted and derelict, desperate
survivors seeking food. Even decades later, much of the western
lowlands of Sumbawa remained knee-deep in ash. Some sold
their children to slave traders to buy rice; others migrated to the
rugged interior of the island where rainfall washed away the ash
more rapidly.[28] The abandoned coastal towns were rebuilt by
slaves from Sulawesi island. The traces of these profound demo-
graphic shifts are still evident today, preserved in toponymy,
dialect and custom. The catastrophe is remembered, too, in
folktales. One, recorded on the Tambora peninsula in 1982,
attributes the eruption to God's wrath after an imam was fed
dog meat during a visit to the island.[29]

While Indonesia was reeling in the months after the eruption
in 1815, across the world people marvelled at spectacular
twilight glows orange-red near the horizon, purple-pink above,

sometimes streaked with dark 'rays'. In the eastern United States the following year, many noted a persistent 'dry fog' that dimmed the sun, and which neither wind nor rain dispersed. No one connected these atmospheric optics with the Tambora eruption, nor did anyone foresee their shocking consequences for food production, which affected the lives of many millions of people, especially in Europe. The dust generated by Tambora had encircled the planet and was diminishing the sunlight reaching the surface – the driver of photosynthesis, on which most life on Earth depends.

To understand these impacts, it makes sense to fast-forward to 1991. That year, a then almost unheard-of volcano, Pinatubo, in the Philippines, surprised the world with a monumental eruption. Like 1815, remarkable sunsets were widely reported around the globe in the following months. Unlike 1815, there were sophisticated sensors on satellites able to quantify what was going on in the atmosphere and on the ground: the eruption opened up a new scientific horizon as it cast unfamiliar light on the earthly skyline.[30]

The most crucial discovery was that the dust girdling the planet was not fine volcanic ash, as had once been thought, but sulphur-rich particles. These had formed in the stratosphere (the layer of the atmosphere that begins around the height that aircraft fly at) from nearly twenty million tonnes of sulphur dioxide gas (that's the weight of ten million cars) emitted during the Pinatubo eruption. The gas had been suffused in the magma deep in the Earth's crust; now it was on the loose in a very oxidising environment – air – and it slowly transformed into specks known as sulphate aerosol. Samples were even collected up in the stratosphere by NASA's high-altitude ER-2 research aircraft (descended from the CIA's U2 spy plane). At massive magnification, they look like wrinkled plant seeds.[31] Owing to their tiny dimensions – somewhere between those of a virus and a pollen grain – these motes stayed aloft for many months before settling out of the atmosphere. More importantly, they were just the right size to reflect a fraction of visible light back into space, cooling the Earth's surface for a few years, with the peak effect in the landmasses of the Northern Hemisphere, the summer after the eruption.[32] You might expect that volcanoes, with burning flames, spewing molten hot lava and searing ash, would heat up the planet, but in fact they do the opposite.

Though several factors, including a volcano's latitude, influence how much an eruption cools the climate, it is the amount of sulphur blasted into the stratosphere that is critical. But how can we ever know how much of the stuff was exhaled by eruptions, like Tambora's, that happened before satellites were monitoring the Earth? It was glaciologists, not volcanologists, who found the answer – in the ice sheets of Greenland and Antarctica.[33] After a particularly large eruption, a little of the sulphurous dust settles imperceptibly over the polar regions and becomes locked in the accumulating ice. A core drilled from the ice can reach back 100,000 years or more in time. Measure the sulphur present in the ice down the length of the core and, like activating invisible ink, a sensational record of past global volcanism materialises.[34] In this way, it is estimated that Tambora emitted three times more sulphur to the stratosphere than Pinatubo. No wonder its effects on climate were profound.[35]

Indeed, they led to the kinds of lived experience that become folklore. In North America, the year 1816, when Tambora's summertime chilling bit hardest, was remembered as '1800 and Froze to Death'.[36] One famed fabulist of the period was Oregonian trapper and mountain guide, Black Harris. In a yarn ascribed to him, he describes weather 'so all-fired cold, it froze icykels on to the star rays, and stopped 'em comin down; and the sun froze so he couldn't shine; and the moon didn't git up at all, *she* didn't'.[37] On the other side of the continent, New

England newspapers, diaries, registers and letters paint a more prosaic picture of the eruption's global ripples.[38] For instance, on 6 June 1816, bricklayers in Bath, New Hampshire, had to down tools because their mortar had iced up. Two days later, eighty-four-year-old Joseph Walker lost his way in the woods of Peacham, Vermont, and spent the night in the open. It cost him a big toe through frostbite. Northerly squalls left nearby Cabot knee-deep in snow, while across the state, farmers who had just sheared their Merinos were tying the fleeces back on to the shivering sheep.

These icy conditions devasted the main staple, corn, as well as the hay crop, leaving one farmer in Rochester, Vermont, pulling the thatch off his roof for fodder.[39] Tens of thousands of other back-country folk gave in, abandoned their farms, and headed for Ohio, Indiana and Illinois in covered wagons or on foot, with whatever goods and chattels they could carry. Many perished on the long journey. The population of Cleveland swelled from a handful to seventy-four in 1817, mostly refugees from New England (briefly among them the father of Rutherford Hayes, nineteenth president of the United States). The exodus left some more upland and remote areas of New England depopulated for a generation, a very distant echo of the abandonments on Sumbawa itself, all the more remarkable since no one then could contrive the connection in circumstances.

It was a similar story the other side of the Atlantic. For instance, on 20 July 1816, the *Times of London* recorded 'very unseasonable weather . . . scarcely remembered by the oldest inhabitant' and foretold that 'the corn will inevitably be laid'. Referring to the aftermath of the Napoleonic wars (concluded at the Battle of Waterloo two months after Tambora exploded), the report added that 'the effects of such a calamity at such a time cannot be otherwise than ruinous to the farmers and even to the people at large'.[40] After a decade of conflict, Western economies were shattered, and countless former combatants jobless and penniless. Things were worse still on the continent. While crossing the volcanic Eifel region in spring 1817, the Prussian general Carl von Clausewitz, famed for writings on warfare, recounted seeing 'ruined figures, scarcely resembling men, prowling around the fields searching for food'.[41] Hikes in grain prices affected millions, and many took to the streets in protest or emigrated.[42] In parts of eastern England, rioters rampaged demanding 'bread or blood' and committed acts of arson and sabotage. The unrest was met with a sometimes lethal response from the authorities.[43]

In the following years, typhus epidemics erupted across Europe. A doctor at Belfast's Fever Hospital estimated 800,000 people were afflicted in Ireland, of whom as many as 65,000 perished from 'the joint ravages of famine, dysentery, and fever'. Astutely, he linked the contagion to malnutrition following the poor harvest of 1816 and the throngs gathering in soup kitchens. Between 1816 and 1817, mortality increased four per cent in France, six per cent in Prussia, and over twenty per cent in Switzerland and Tuscany, where many starved to death. In Europe, 1817 became remembered as 'the year of beggars'.[44] Unsurprisingly, the famine only accentuated inequalities and political tensions. Anti-Semitic protests erupted in the Bavarian city of Würzburg in August 1819, the violence quickly spreading to Copenhagen, Amsterdam and Krakow.[45] While New Englanders were heading west in the United States, tens of thousands of Irish men and women were emigrating to North America (pre-empting what hundreds of thousands would do during the Great Famine thirty years later). The conditions aboard ship during the Atlantic crossing were hardly conducive to quelling the spread of typhus, and many were found sick on arrival and placed in quarantine.

Tambora has been implicated in many other manifestations, trends and innovations of the period, including cholera outbreaks in India, opium addiction, the writing of *Frankenstein*, the blossoming of the art songs of Beethoven and Schubert, J. M. W. Turner's intense sunsets, three decades of mostly futile, often lethal pursuit of the Northwest Passage, and even the invention of the bicycle. But the problem with any claim of causality is the implied counterfactual. Can we say that if Tambora had not erupted when it did, Mary Shelley would not have penned her Gothic masterpiece, the farmers of Yunnan province would not have switched from rice to poppy cultivation and Sir John Franklin might have enjoyed a long life on his naval pension? In the case of Tambora, the links to climatic change and thereby to agronomic impacts in Europe and North America are convincing. But any knock-on consequences must surely have a great deal to do with the eruption's post-war timing, and the extent to which authorities and institutions acted wisely to alleviate deprivation. When contemporary observers themselves drew connections – as they did in the years after 1815 – between weather, crop yields, uprisings, public health and disease, this, for me, strengthens our interpretations of affairs.[46]

★

As a volcanologist, I am often asked: 'Isn't Yellowstone overdue for "the big one"?' Before the BBC's 2005 film *Supervolcano*, which envisioned the reawakening of Yellowstone, you would only have had the misfortune to hear the expression 'supereruption' at the dentist. Thanks to the TV drama, however, volcanologists quickly appropriated the term, later defining it as an eruption producing 1,000 cubic kilometres or more of pumice and ash. It is not easy to picture such a volume of rock, but it would be equivalent to burying all of Greater London to double the height of its tallest skyscraper (the Shard), all of France surpassing head height, or the entire United States well above the ankles.

Yellowstone's last super-eruption took place 640,000 years ago, long before our species emerged. More interesting to consider, I would argue, is Sumatra's 'Youngest Toba Tuff' eruption, just 74,000 years ago. By that time, our ancestors were using advanced stone tool technologies, and likely knew how to tell a good yarn, too. This was a blast 150 times bigger still than Tambora's, disgorging enough pyroclastic rock to cover the whole of the United States to the depth of a one-storey home. About a third of the deposit piled up on northern Sumatra, and much of the rest lies beneath the floor of the Indian Ocean.[47] Given its great scale and period, close to both the onset of the last Ice Age and the time when *Homo sapiens* dispersed out of Africa, this 'super-eruption' has become entwined in debates concerning climate change and human prehistory.

The clearest topographic trace of the super-eruption is an elliptical crater lake, sixty miles long, amidst the peaks, forests and rice terraces of the Batak region of northern Sumatra, an area explored extensively by Franz Junghuhn in the early 1840s.[48] The caldera is so vast that from the ground it is hard to get the sense of being on a volcano – the scalloped rim and blue-grey water simply dissolve in haze and skylight far short of the distant margin. Pumice deposits from the eruption dazzle in canyon walls and extend deep below ground, but perhaps more exceptional is the mostly unseen veneer of dust that mantled a fifth of the Earth's surface.

While there are only minor quibbles about the quantity of pumice and ash involved in this cataclysm, there is no consensus on how much sulphur it released into the atmosphere – the estimates are more like guesses and vary enormously. Some sulphur layers in the polar ice cores have been suggested as possible candidates, but none has yet been definitively attributed to Toba. This hasn't prevented climate scientists from running computer models to gauge the super-eruption's global repercussions – they are interesting in their own right, but none can be claimed to provide a reliable picture of what actually happened until we properly constrain Toba's sulphur potency.[49] However, even the most extreme model scenarios investigated do not simulate widespread glaciation, so one thing that's clear is that Toba did not trigger the last Ice Age.

Toba was on my gap year tour before going to university – I stayed for several days on Samosir island (an uplifted block of the 74,000-year-old ignimbrite rising from the centre of the crater lake). My temporary home was a Batak house with traditional saddle-shaped roof, surrounded by plots of rice, cassava and vegetables, and a grazing buffalo or two. I ate my first avocado there (this really was a journey of self-discovery – I took to them at once and consumed around four a day), drank the local liquor (made from sugar palm sap), coughed my way through a homespun cigar and swam in the placid lake water, oblivious then, in my pre-volcanologist naivety, of the immense amount of magma below me.[50] But despite this early acquaintance with Toba, I ended up studying the eruption more closely by going further away – halfway across the Indian Ocean to southern India.

The site lay between a dry riverbed and the village of Jwalapuram in Andhra Pradesh. A local cottage industry had sprung up to mine a layer of ash just below the surface. It was sold as an abrasive for use in detergents. There are no volcanoes anywhere nearby, and its chemical make-up is an exact match for Toba: this is the fallout from the pall of fine ash carried by stratospheric winds across the Indian Ocean from Sumatra.

Even more exciting, the ancient soil layers sandwiching the thick bed of ash contained many prehistoric tools: flakes, scrapers and cores made from chert, chalcedony, quartz and limestone. I was working in a team with archaeologists, and while they painstakingly recovered every piece of worked stone, I used DIY-store trowels, knives and paintbrushes to reveal subtle variations in the deposits and to exhume the soil surface on which they rested. The stifling heat, humidity and dust in the pits were overwhelming. Biting flies tormented me. But these discomforts were eclipsed by the thrill of exposing the moment the ancestors witnessed darkness at noon and the earth turned to powder. While my typical fieldwork – pointing spectrometers at drifting gas clouds – addresses the here and now of volcanic action, revealing terrain buried for 74,000 years felt like time travel.

The sugary deposits preserved remarkable details, such as the tunnels through which bugs had escaped their rude burial in fallout. I also found a level in the ash with abundant leaf impressions – I guessed that the trees were defoliated by the veneer of dust. When the sky cleared 74,000 years ago, the toolmakers and hunter-gatherers must have gazed in horror at the infinite carpet of blinding white powder; perhaps, some-where, their footprints are preserved in it. Above this layer were several much thicker stripes of ash with tell-tale signs of mud cracks. This matter must have settled out on the surround-ing hills but was then washed downslope by monsoon rains. If so, that would mean the eruption did not greatly disrupt the cycle of wet and dry seasons, as some have suggested. The deluge of wet ash had set like concrete around tree trunks and branches, helping to petrify them. It's hard to imagine human populations continuing to inhabit such shifting ground. Perhaps they left their formerly wooded homeland and sought resources and cave shelters on higher ground. What stories did their descendants tell of dislocation and survival? Certainly none are still preserved over such a span of time, but is it possible that ancestral experiences of such crises helped to shape what we call 'human instinct'?

We know that humans were in India when the ash fell because of all the implements, but two key questions remain: what species were they and did they endure the catastrophe? Since there are no human fossils at this site nor of this period from anywhere on the subcontinent, the only clues are the stone tools. Unfortunately, it is not easy to attribute one flake or scraper to Neanderthal manufacture and another to *Homo sapiens*. After much measurement and characterisation, the specialists concluded they most resembled tools found with *Homo sapiens* fossils in southern Africa, and so credited the Indian samples to the handiwork of our species. If correct, this implies 'we' had reached the subcontinent more than 74,000 years ago. But others disagree.[51] Resolving the matter would have profound implications for understanding the drivers of migration of our species from Africa to Asia and beyond, as well as our encounters with other extant humans, including Neanderthals and Denisovans.

But might the eruption mean more than a convenient marker pen for archaeological dating? Could it have chilled global climate enough to influence the human trajectory? In 1998, the anthropologist Stanley Ambrose proposed that Toba's paroxysm sparked worldwide environmental devastation on the scale of a 'nuclear winter', the Armageddon some consider might follow nuclear warfare. Ambrose reckoned that Toba's climate change almost eradicated our ancestors.[52] Though his argument inspired and provoked a great deal of debate and further study, it has not been substantiated. Indeed, the evidence emerging from archaeological studies and analyses of ancient environments in Africa, in places where the Toba ash has been found in sediments, argue against universal and lasting depletion of water, food or shelter.[53] Despite its great magnitude, the Toba super-eruption may have

released comparatively little sulphur into the atmosphere, limit-
ing any global climatic reverberations.

The case of Toba urges us to think forwards, too: will the
next super-eruption pose an existential threat? Will Yellowstone annihilate us all one day? The chances of such extreme
volcanism are not so remote as one might think. The last
example that we know of convulsed the North Island of New
Zealand (Aotearoa) 25,000 years ago, and the global return
period of such large events looks to be as short as 17,000
years.[54] That equates to around a 1 in 200 chance there will
be another before this century is out. But the likelihood
Yellowstone will be next in the frame is remote. What's more,
the popular trope that the Wyoming hotspot is running behind
schedule stems from a misreading of its super-eruption score
card: past events occurred 2.1 million, 1.3 million and 640,000
years ago. Somehow, this has been taken to imply 600,000-year
intervals. Even taking the average of the two intervals between
events (730,000 years) is unhelpful, as there is no reason to
expect Yellowstone to behave like a clock. Volcanologists with
the United States Geological Survey regard the likelihood of
another large caldera-forming eruption of the volcano as 'below
the threshold of useful calculation'.[55]

Lake Toba remains as tranquil as it was when I saw it as a
nineteen-year-old backpacker, but a deadly eruption of nearby
Sinabung volcano has recently tapped the super-volcano's
magma store.[56] Until it reawakened, sulphur was still being
mined from its fuming craters. In this case, no folk tales or oral
tradition hinted at ancient calamity – the last eruptions presumably occurred beyond the reach of memorialisation in myth.
Nor did science prepare the population – no sensors were tracking the volcano's pulse before it kindled. Even the eagle-eyed
Junghuhn neglected to mention Sinabung in his travels in Batak
country. So its revival in August 2010 (just two months before

Merapi's paroxysm on Java) took the communities at the foot of the volcano completely by surprise, the more so since the first blasts occurred at night as people slept.

It is possible the magnitude 9.3 'Boxing Day' earthquake in 2004 (which triggered the catastrophic Indian Ocean tsunami) aggravated Sinabung's magma reservoir, but we don't really know what roused the volcano. The activity picked up in 2013, prompting the authorities to declare an exclusion zone around the foot of the juddering mountain. Its boundary was porous, though, and sixteen people were killed a few months later by *nuées ardentes*, among them several climbers who were on the mountain despite the warnings.

I visited Sinabung with a film crew in 2016. With permits from the national geohazards agency, we drove towards the volcano through empty villages. Pullulating weeds festooned walls and homes, and had colonised the ash-plastered road, jumping at their chance to overwrite brick, asphalt, galvanised metal and satellite TV antennae. But stretched across the yard beside one house, I spotted a washing line hung with recently laundered clothes. In this mundane detail, I understood the challenge of protecting people from volcanic disaster – even when the alarm has sounded, even when neighbours have perished, it is very, very hard to leave home.

We stopped at the edge of the village, where *nuées ardentes* had scorched a small clapboard home. The metal roof had corroded away; a singed soft toy lay on the ground, half-concealed by bindweed; a jacket hung from a hook – signs of intense but momentary heat as the *nuées* raced through. Looking up to the volcano instantly took me back twenty years to Soufrière Hills on Montserrat. In fact, the resemblance was uncanny: rock crumbling off the lava dome at the top and clattering down a pink-grey scar in the volcano's side; the once-verdant flanks bleached by acid fumes; former roads and farm

terraces now just ghosts in the topography. I felt, too, the same
sense of immediate menace, and kept a very close eye on the
chunks of lava falling off the dome. In another echo of
Montserrat's long-lived crisis, Sinabung's crater continues to
simmer and spit, to this day.

My encounters with Merapi, Tambora and Toba, and collaborations with historians, archaeologists and geographers, have
brought home to me that a wide range of conditions – socioeco-
nomic, political, cultural, environmental, chance – influence
human agency, whether at individual, community or global
scale. History is *story* (the shared etymology of the words is no
accident) and storytelling is an essence of humanity. Folklore,
myths and stories help us make sense of the world, and of our
past, present and future. Though they are often set apart, the
indigenous or local knowledge of communities can go hand in
hand with science, something that Junghuhn understood.
Narrative pervades science as much as it does fable, albeit its
rhetoric is more dialectic than lyrical. Good science can raise
the level of historical arguments, and society could do more to
value the public and educational value of interdisciplinarity.

Nowhere have I more profoundly experienced the magical
side of volcanoes than in Indonesia. It's where I realised how
climbing a volcano differs from scaling a regular mountain –
there is always a mystical nature you cannot discover until you
reach the crater and peer in. I once scrambled up the highest
and holiest mountain in Java, Semeru. According to tradition it
is formed from the peak of Mount Mahāmeru, sacred in Hindu
cosmology, and was brought to Java to stop the island from
shaking. While I watched from a safe distance, a sagging dome
of lava exploded every ten minutes or so, launching grey-brown
cauliflower clouds that coiled into the sky. This was impressive
enough, but the accompanying noise was unlike anything else I
have ever heard. It was a siren, a blizzard and a wailing child all

at once – an animal, really – and it was followed by a sonorous chugging, like a steam train gathering speed. The timbres and tones and resonances were incredible and moving. Later, an Indonesian colleague pointed to a saddle near the summit of the neighbouring volcano and said it was known locally as 'the night market of the ghosts' where spirits congregate, dine and gossip. Thinking of those unworldly sounds, recalling, too, the ethereal blue light of burning sulphur at Kawah Ijen, and nocturnal glow of Bromo, it strikes me that volcanoes must have always been wellsprings for the imaginary.

White Mountain, Heaven Lake

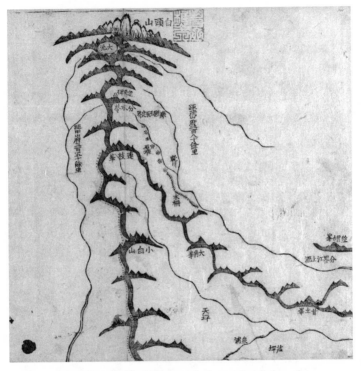

Paektu, the 'white-headed mountain' and
source of the 'dragon's vein'.[1]

'We greet the sun at Paektu, and greet the unification at Halla.'
– Ri Sol-Ju, Reuters[2]

Hands clasped and held aloft, Moon Jae-in and Kim Jong Un smile for the camera in front of Heaven Lake atop Mount Paektu. The 'first ladies' beside them are clapping. It is a picture of celebration and ceremony. But why choose a crater formed by one of history's largest eruptions for such a momentous reunion of the leaders of North and South Korea, two countries technically still at war? Was it for the joy of wordplay, a summit at the summit? Or the frisson of an earth-shaking encounter where the Earth shook?[3]

The choice of backdrop for this historic moment in 2018 comes down to the fact that this volcano is the source and soul of the Korean people. Mount Paektu, it is said, radiates the spirit that will one day reunite Korea. The volcano has become a symbol: it emanates revolutionary spirit; its name is sung in kindergartens and schools; its leaders draw their legitimacy by posing beside its crater; it fills the dreams of a divided people.

Paektu is also well known to volcanologists, thanks to a notable Japanese expert on ash deposits, Hiroshi Machida. But his work, in the 1980s, started neither in the Democratic People's Republic of Korea (DPRK), nor in China – the two countries that share the territory of the volcano – but far away in northern Honshu, Japan, where he found two prominent bands of ash amidst fossiliferous sediments.[4] At first, he thought they must have come from the nearby Towada volcano. But chemical analysis revealed that the upper layer, as thick as a couple of paperbacks, did not match any of Japan's volcanoes. The ash was also present on Japan's northernmost island, Hokkaido, and here what startled Machida was that its depth was the same: this could only mean the volcano was very distant.[5]

He turned next to sediment cores from the seabed west of Japan, and there, once more, was the mysterious ash – but now it was thicker. This clue led to a 1940s geological survey of Mount Paektu, which mentioned deep pumice deposits and a buried forest near the volcano's peak. The report included chemical data for the rocks – they matched the ash in Japan! Machida had solved the puzzle, and estimated that the eruption took place around 1000 CE, based on historical dates for other ash layers in the sediments. Ever since, it has been known as the 'Millennium Eruption'.[6] Subsequent work on the distribution of ash and pumice ranks it among the two or three largest eruptions of the past two millennia.

Mount Paektu (meaning 'white-headed mountain' in Korean) is only one of the volcano's names, which at once speaks to its location in a region long contested by different polities and empires; it is Golmin Sanggiyan Alin ('long smoky mountain') to Manchurians, and Changbaishan ('long-white mountain') to the Chinese.[7] For Koreans, it is the site where King Tang'un, the progeny of gods and a she-bear, was said to have been born in 2333 BCE. According to a thirteenth-century Buddhist source, King Tang'un established the first Korean nation and ruled for 1,500 years. The volcano's importance was more spiritual than political, however, during the period of the powerful Koryŏ kingdom (918–1392 CE). Mount Paektu was then considered the wellspring of the Earth's invisible inner energy, *p'ungsu*, which was dispersed through the 'dragon's vein' of mountain ridges that hold the Korean peninsula together like vertebrae. The mountain spirits, called *sansin*, who took the form of benevolent old men, were venerated at shamanic shrines and Buddhist and Daoist temples.[8]

While spiritualists saw a magical realm within the rocks, emperors had more political concerns above ground, not least concerning where the frontier – such an imaginary construct – should run between Manchuria and Korea. In 1712, a border commission of sorts demarcated it along the courses of two rivers, the Yalu (the Amrok in Korea), which originates as a waterfall high up on Mount Paektu, and the Tumen, thereby sharing the volcano between the two empires. The Manchu emperor Quianlong would write later: 'To ascend to the primitive source of our August Race . . . we must carry ourselves to that mountain, distinguished in like fashion for the size and for the colour with which it shines.'[9] Everyone in the region, it seems, had an ancestor story tied to the volcano.

The 1712 treaty might have settled the question of where one state ended and another began, but the position of the upper reach of the Tumen proved ambiguous and fresh negotiations took place in 1887. At this time, the Romanovs and Japan were also taking an interest in the region. In 1904, after the Russo-Japanese War, Korea came under the control of Tokyo. Five years later, in 1909, Japan ceded the summit of Paektu to China in a border agreement. This stoked Korean nationalism and the old myths surrounding the volcano were revived and repurposed. King Tang'un was promoted as a historical figure; a calendar rooted in his supposed birthdate was proposed, and festivities staged to mark his anniversary. A new religion of 'the divine progenitor' even sprang up. Since Paektu was Tang'un's birthplace, the volcano itself became a sacred symbol for Koreans. A small stele said to date from this period stands at the crater's edge on the DPRK side of the volcano. It is inscribed with a text calling on the netherworld spirits of Heaven Lake to liberate Korea and bring prosperity and happiness to the people.

That appeal was answered in the form of Kim Il Sung, who, in the 1930s and 1940s, launched offensives against Japanese forces from 'secret camps' on Paektu's slopes. His son, Kim Jong Il, was, by repute, born in one of them. In 1962, Mao Zedong seemingly accepted that the DPRK had been wronged in the 1909 boundary negotiations, and signed a new treaty that put the frontier through the centre of the crater.[10] The mountain became a beacon of the struggle against imperialism, and the founding of the nation, by extension linking the Kim family irrevocably to both King Tang'un and the volcano; out of revolution, the 'Paektu bloodline' was forged.

This notion was reinforced in 1993, when North Korean archaeologists claimed to have found Tang'un's tomb. An elaborate mausoleum to house his bones was built on a hill half an hour's drive from Pyongyang. More than ever, Mount Paektu became a sacred ancestral site of pilgrimage for North Koreans. Meanwhile, South Koreans could only visit the mountain from the Chinese side – as a geographical reminder of the division of the peninsula, Paektu came to symbolise the dream of reunification. It is hardly surprising the volcano's name is sung in the national anthems on both sides of the Demilitarised Zone that divides North from South.

*

As the site of one of the largest eruptions in history, Mount Paektu had been on my radar for some time,[11] but I assumed the sensitive border region crossing the volcano was off limits for foreigners, and had never contemplated going there. Strangely, it was the massive 'Tōhoku' earthquake in Japan in March 2011 that opened the door to my visit. The temblor generated a powerful tsunami and claimed thousands of lives in Japan. Further shockwaves were felt around the world as the shaking and flooding led to reactor meltdowns at the Fukushima nuclear power plant, releasing radioactive materials into the environment. Some seismologists in China suggested the earthquake was so powerful it could have disturbed the magma beneath Paektu volcano.[12] The Chinese had also been investing in volcano surveillance ever since thousands of tiny tremors jangled the mountain between 2002 and 2005.[13] These developments concerned the North Korean geoscientists, who worried that their volcano-monitoring equipment wasn't up to scratch.[14]

It was a friend and science journalist with the American Association for Advancement of Science, Rich Stone, who first told me of Pyongyang's interest in hosting a volcanologist.[15] In turn, he'd got wind of the North Koreans' wishes through Kosima Liu, co-director of an NGO in Beijing, where Rich was then based. Rich and Kosima would be central to the whole enterprise. Meanwhile, I contacted James Hammond, a specialist in finding magma underground using earthquake data. His experience would be invaluable, but I also knew that if anyone would agree to join me on a trip to Pyongyang with a few weeks' notice, it would be James.

We started making arrangements, but hit an immediate problem – we couldn't contact our counterparts in Pyongyang. I'd had to overcome all kinds of logistical challenges with fieldwork elsewhere in the world – trouble with equipment, supplies, transport, permits – but one thing I'd always taken for granted was being able to email, fax or phone my collaborators. In our dealings with the DPRK, however, I had to communicate through a chain of intermediaries, starting with Kosima, who was, in turn, in touch with Ryu Kum-Ran, who works for an agency called PIINTEC. She would then relay messages to and from the scientists in Pyongyang.[16] It was through this ping-pong exchange that we planned our trip.[17]

We arrived in the North Korean capital full of expectation, but I had not anticipated seeing Kim Jong Un, as well as his father and grandfather, all within twenty-four hours. It was 9 September 2011, the sixty-third anniversary of the founding of the DPRK, and there was to be a military parade in Kim Il Sung Square. We found ourselves on the guest list.

The human and hardware superabundance was eye-popping, and the martial music clamorous. At the end of the processions of goosestepping soldiers at arms, and motorcades of combat vehicles and rocket launchers, 100,000 human pixels generated vast collective mosaics like a gigantic TV screen. The thunderous presentation of coloured placards involved synchronisation to the millisecond, presenting image after image of revolutionary slogans, characters and scenes. Mount Paektu was a recurring feature. At the end, to a reverberating roar of cheering party members, Kim Jong Il and Kim Jong Un stood at their balcony overlooking the raked stands of spectators and waved in our general direction. A little later, we were driven to the mausoleum of the Palace of the Sun for an audience with the embalmed remains of Kim Il Sung, the founder of the DPRK.

Everywhere we went in Pyongyang, Mount Paektu confronted us – it featured in propaganda posters, in effigy atop the Arch of Triumph (a fraction taller than the Arc de Triomphe in Paris), in a giant mosaic behind the statue of Kim Il Sung at Mansu Hill, on food labels and on murals depicting the Eternal Leader, Kim Il Sung, and his son and successor, Kim Jong Il, on a visit to the crater rim. It was dawning on me that the volcano was much more than a physicality, and that in encountering all three Kim leaders, I had come face to face with the Paektu bloodline.

Interspersed with our tour of sites and monuments of Pyongyang, we met our new colleagues, all of whom wore loyalty badges depicting the DPRK flag or the image of Kim Il Sung. It felt claustrophobic not to be able to step out into the city, justly famed for its utopian architecture, and wander freely, but we understood implicitly that to do so would be to violate a trust.

We flew to Samjiyon, near the foot of the volcano, in a Soviet-era Antonov. Apparently, the plane was given to Kim Il Sung by Nikita Kruschev, so it wasn't the latest model, but it got us there in style. A farm tractor pulled a trailer with our luggage as we waited outside the airport building. Our first stop was another grand monument beside Lake Samji, dominated by a commanding bronze of a young Kim Il Sung clasping binoculars.[18] In military dress, with jodhpurs and boots, he stood on a white granite plinth in the form of Paektu, whose actual peaks were visible in the distance. We then settled into our base for the next days, the Bae Gae Bong Hotel, set in a clearing in larch forest.

Nourished on noodles, kimchee and boiled fern, we set off the next morning, driving through Samjiyon, a pretty town with steep-roofed houses. Pedestrians wearing combat fatigues, flat caps and rubber boots filed along the pavement carrying bulging cloth knapsacks. Above them, posters depicted workers in hard hats, their fists clenched in revolutionary fervour. Once out of town, we followed a corrugated dirt road into a dense forest of Korean pines, ash and oak. There was little traffic, save for army trucks bringing pilgrims to and from the crater, though we saw many people emerge on foot from the woods toting buckets of blueberries or baskets full of yellow mushrooms. Here and there in clearings were cabbage patches or scurrying piglets. Were it not for the pumice spilling into the roadside where disturbed by tree roots, it would have been easy to forget in these dense woodlands that we were climbing a volcano.

After an hour on the road, we stopped at a volcano observatory near an eroded volcanic cone, Mudu Peak. The director of the Volcano Institute, Kim Myong Sung, welcomed us and, while showing us seismographs and geochemical equipment, said we were the first foreign visitors he'd ever received. He explained, through our interpreter, that there are six satellite observatory posts on the mountain. 'But our power supply can be interrupted and there are uncertainties in our measurements,' he admitted, 'so we welcome more international cooperation to improve our monitoring effort.' His colleague, Professor Kim, described the seven stages of an alert system used to characterise the level of volcanic threat. 'We are at stage three, which means there is uplift of the volcano and anomalies in hot springs.' I asked, apprehensively, what stage seven indicated. 'A full-blown disaster,' he replied. 'But we do not see any worrying signs now.'

It soon became clear that our hosts were hoping we might help to equip their field stations as well as spark international collaboration after working for so long in isolation. Professor Kim noted that there was a great deal of interest in finding out exactly when the Millennium Eruption had occurred, and whether there had been any subsequent activity on the mountain. As a final word before we left, he reminded us: 'Mount Paektu represents the soul of our nation and our ancestral home. You should have a strong heart to climb it.'

In fact, the ascent required no exertion at all on our part, since the road led all the way to the top. Before long, the coniferous forest gave out in favour of a thin cover of beech and birch trees, already tinged with autumn colours. Then, ten minutes later, we were truly in the open, crossing an alpine tundra, the pumice-cloaked slopes ahead rising to an undulating ridgeline that marked the caldera's edge. Anyone anticipating a scene like Mount Fuji, with flowering cherries and concave slopes steepening to a pinnacle, would be disappointed. The Silk Road explorer and mystic, Francis Younghusband, who climbed Paektu in 1886, was among those underwhelmed by the topography: 'I cannot say . . . it inspired us with awe commensurate with the mystery ... attached to it ... it was not the snow-clad monarch we had expected to see.'[19]

The plant life, though, was extraordinary: creeping rhododendrons, bilberries, moss and open grasslands formed a gorgeous patchwork of russet, crimson, umber, rust, pale gold and bronze in the afternoon sunshine.[20] Despite this profusion, there was nothing you could really call a soil – just slowly weathering pumice. The only living trees were stunted and twisted, yet along the roadside we saw many thick trunks poking out of the pumice. We realised these were all that remained of a verdant forest extinguished and entombed by the Millennium Eruption. A little excavation around the stems revealed intact bark – it was hard to believe these trees were so old. I tried to imagine the forest that had once thrived here, and thought how gradual the pace of ecological recovery has been compared with what I had seen on tropical volcanoes, where as soon as it rains, rampant vegetation rebounds. Even with the passage of a thousand years, this place looked nothing like it must have done before the eruption.

Ecologists have learned a lot about how life reboots after major disturbances by studying developments after large volcanic eruptions, which instantly sterilise a landscape. Recovery turns out to be a complex process dependent on interwoven aspects of climate, ecology and landscape, as well as chance events. One survival lesson from the 1980 eruption of Mount St Helens in Washington state, U.S.A., was that being small was good. Burrowing animals fared best of all, and gophers were the star regenerators. Not only did they withstand the violent pyroclastic blasts below ground, their industrious tunnelling mixed buried soil and pumice and ash, bringing nutrients, seeds and spores back to the surface. But a thick layer of pumice does more than displace flora and fauna; it changes the shape of the land, and the way water runs off it. Lakes, rivers, and coastal waters can all be choked by the quantities of sediment mobilised during heavy rains.[21]

Re-establishment of the ecology of Krakatau in Indonesia after its 1883 eruption was also rapid, thanks to seabirds and ocean currents and the biodiverse shores of Java and Sumatra. Within a year, a solitary spider had found its way to the remnants of the island volcano, and within a decade, dozens of plant species were re-established. Twenty-five years later, 200 animal species, mostly insects, were thriving, and as time went on, the forest matured.[22]

In the case of Paektu, several variables hindered recovery. First, the great thickness of pumice permanently sealed the seedbanks in the original soil. Second, the summer growing season is short, and the summit area remains under snow for up to eight months of the year; the annual mean temperature is well below freezing, and most of the precipitation falls as snow or hail. And third, the pumice is nutrient-poor, so porous it does not hold water well, and constantly shifting in the swirling winds that scour the mountain. This is why the summit area today looks more like the Arctic than the densely forested land that was once there.

The last section of road up to the Paektu summit negotiated the crater rampart by several switchbacks. We were dropped off at the foot of Janggun Peak, the highest point on the mountain, and escorted into a small building, which we realised was a cable car station. Our main contact, Ryu Kum-Ran, ushered us into one of the red cabins and we waited for the mechanism to fire up. But she shortly reappeared, popped the door open and advised: 'It is better to walk. There are more than 2,000 steps, so it is hard, but there could be a power cut and we cannot guarantee your life in the gondola.' This was persuasive, though the zigzagging aerial staircase of hefty granite treads on scaffolding anchored in the steep inner caldera wall also proved intimidating where the handrails had broken or where steps had slipped through the girders to crash on to the boulders below.

If the flanks of the volcano seemed unimposing, the interior, forged in the convulsions of the Millennium Eruption, evoked all the drama of Korean history. The pinnacles and serrated ridge of the crater rim from this aspect were sheer-faced bluffs plunging into sweeps of coppery, slabby stone and brown and grey scree. The streaks of colour were softly reflected in Lake Chon – Heaven Lake – a stunning counterpoint to the rucked, scarred, sculpted cliffs and raw rock strata. Even Francis Younghusband had been impressed:

> At last we reached the saddle, and then, instead of the pano-
> rama we had expected, we looked down in astonishment on a
> most beautiful lake . . . held in on every side by rugged precip-
> itous cliffs . . . Like a sapphire in a setting of rock.

We passed groups of uniformed Korean People's Army soldiers on their way up, their expressions turning from exhaustion to curiosity at the sight of us. Reaching the lake shore, following the cue from our Korean colleagues, we removed shoes, rolled up trousers and stepped into the limpid, icy water in a gesture of reverence to the sacred ground and the *sansin*.

Our hosts then led us through a doorway in a bunker housing an austere outpost of the Volcano Institute. Remarkably, the small observatory is manned year-round – the dedicated crew of four is cut off for months through the long, harsh winter. They might as well have been in Antarctica. A slogan painted on the wall read: 'Great Leader Kim Il Sung lives together with us', and I couldn't help thinking, as we sat in a circle on the floor to eat packed lunches, *so does the magma, probably not so far down.* The observatory chief outlined the monitoring effort, which included regular boat trips across the lake to sample hot springs and underwater fumaroles for chemical analysis. 'When the lake freezes over,' he explained, 'we go on foot.'

Back at the crater rim, cloud was swirling around its crags and crowns. The sun was low and turning the grasslands on the serried slopes to a burnished copper. In the next days, we visited more sites of cultural interest, another of the observatories, and discussed what we had learned from each other. We shared thoughts on research priorities, and James and I pledged to seek funding to nourish our exciting new collaboration. The key scientific questions that emerged were: is magma present beneath the volcano, and if so, where? And when *did* the Millennium Eruption happen?

*

It took us two years to fulfil our promise of further cooperation, and much advocacy from supporters in Washington DC, London and Beijing, as well as Pyongyang. Among our experienced backers was Norm Neureiter, a senior advisor to the American Association for the Advancement of Science. Norm, a formidable polyglot with a long career in industry and government service, had worked in the White House Office during Nixon's presidency as a go-between with the Soviet Union and China on scientific cooperation. This Cold War-era experience convinced him of the possibilities for scientific collaboration to provide new room for action, even when official diplomatic and political channels are severed. For instance, dialogue between individual scientists is often credited with paving the way for the meetings between presidents Reagan and Gorbachev that led to nuclear arms control between the USA and USSR. Norm helped us get funding, while James became a shuttle diplomat on top of his day job as geophysicist, travelling to Berlin, Paris and Beijing to lay the groundwork for our ambition, a two-year deployment of seis-mometers and geological fieldwork on Mount Paektu.

A key issue we had to grapple with was compliance with the strict sanctions imposed on the DPRK. Anything we might bring into the country – hardware, software, even the know-how in our heads – had to conform to stringent international regulations. Some of our plans had to be shelved because the associated kit was considered 'dual use' – that is, it could conceivably be used for military purposes. In particular, we wanted to bring equipment for measuring electrical and magnetic fields in the earth, but some of its components could, in principle, be used for submarine detection. Another headache for us was money – too much of it (for a change). There are no international banking arrangements with the DPRK, so we had to travel around with anxiety-inducing quantities of euros. Finally, with all our export licences approved, and wads of cash in our pockets, we returned to Pyongyang on 1 August 2013.

This time our team included Kayla Iacovino, who then worked at the U.S. Geological Survey. I had supervised her PhD studies on Mount Erebus in Antarctica, which has some geological affinities with Paektu. Her expertise would bring a new dimension to our enterprise. On the PIINTEC side, we were joined, too, by the boyish and jovial Mr Kim, who proved to be an accomplished accordion player and sedulous escort. Kim Jong Il had died at the end of 2011, and now there were two towering bronzes at Mansu Hill, father and son side by side, and two portraits on the loyalty badges.

As an American citizen on her first visit to DPRK, Kayla was informed in detail by our hosts about North Korea's tenets of self-reliance and 'military-first' policy, of the imperialist aggression of the U.S., atrocities committed during the Korean War, the hostile 'war games' going on south of the Demilitarised Zone, and what North Koreans regard as a U.S.-led sanctions regime aimed to cripple them economically. She handled, with aplomb, both the lecturing and the general air of surprise exhibited by our DPRK colleagues at finding themselves in the company of an independent female scientist.

The six seismometers, which had been separately airfreighted, arrived safely. This was a great relief, especially for James, but the most symbolic moment for us was to sit around the table with the PIINTEC director to sign the memorandum of understanding that laid out the rules of engagement of our 'Mount Paektu Geoscientific Group'. Spearheaded by James, it had been teased out through tireless negotiations and amendments, and was the formal centrepiece of our ground-breaking alliance. A few days (and several surreal karaoke evenings) later, we were back in the by-now familiar surroundings of the Bae Gae Bong Hotel in Samjiyon.

Professor Kim joined us again, along with more than a dozen of his colleagues. Among them were Dr Kim Ju Song, an engineering geologist, and Dr Ri Kuk Hun, a geophysicist and remote-sensing specialist with the State Academy of Sciences. Both men were in their forties.[23] Dr Ri had a canny expression and a twinkle in his eye; his khaki shirt was invariably unbuttoned, revealing a well-filled, low-necked white vest. The leaner and well-groomed Dr Kim looked more intent. Many of our Korean scientific colleagues smoked a great deal, indoors and out, and at meal times would knock back quantities of *soju*, a strong distilled spirit, and sometimes an equally potent brew euphemistically labelled as blueberry wine.

Kayla and I accompanied Dr Kim on hikes to his favourite rock exposures. He had studied the volcano for twelve years, spending most of his summers on the mountain, searching for clues to its eruptive history. It was a month earlier in the year than my previous trip, and Paektu's alpine flora were in full summer bloom: delicate yellow Arctic poppies, aquilegia and many varieties of herb abounded. Since the Millennium Eruption, there has been limited human disturbance, and no grazing high up the mountain, so this is quite an untouched ecosystem.[24] My joy at being back on the mountain was accentuated by the contrast between hotel confinement or escorted visits, and comparatively liberal rambling across Paektu's blossoming wilderness.

The most exciting site Dr Kim showed us was located in a small valley cutting into the Millennium Eruption deposits. In general, ash and pumice beds stay in place after the dust settles, so the basic law of stratigraphy applies: oldest rocks at the bottom, youngest at the top. Here, a blanket of black pumice and ash was draped on a very thick bed of white pumice, which we had seen for miles around. Whereas the pumice fragments in the white unit were sharp-edged, the fist-sized lumps of black pumice were rounded. These rocks were all from a single eruptive episode likely to have played out over a period of a few days. As the eruption progressed, it tapped a deeper and chemically distinct magma. (A very similar transition is seen in the deposits of the 79 CE eruption of Vesuvius.) The angular fragments indicate fallout from soaring ash clouds, but at some point the eruption column collapsed, generating ground-hugging *nuées ardentes*, whose violent and chaotic internal collisions abraded their rocky cargo.

Dr Kim led us next into a short tributary and pointed excitedly at the base of the white pumice, which lay on brown clay. 'You are the first people I have shown this site,' he said, proudly. Here was an amazing geological time capsule like the one I had studied in India, where the Toba ash covered ancient soil. Everything directly beneath the pumice had been a thriving habitat; now it represented a 'kill layer' of organic debris. Meanwhile, the ash immediately above offered clues to the opening salvo of the eruption. I excavated carefully to unveil more of the boundary, exposing a fallen birch tree. Its trunk was scorched on one side and had toppled in the direction away from the crater. These were hallmarks of the kind of violent horizontal blast of hot ash that had knocked down and seared trees during the 1980 eruption of Mount St. Helens.

Over several days, we scoured numerous valleys and peaks, and collected more than a hundred rock samples. As we looked out across Heaven Lake from the crater rim one afternoon, Dr Kim stared into the future: 'There are 200 million tonnes of water down there. What will happen to it when the next eruption happens?'

Communication was easy enough through our interpreter, but much was still lost in translation. Owing to North Korea's isolation, Dr Kim and his colleagues lacked exposure to the cutting edge of volcanology. They had little more than a hammer and a microscope to study the rocks, whereas Kayla could subject samples to penetrating micro-analysis with state-of-the-art lab instruments. Beyond the technical mismatch, international volcanology is, today, a melting pot of natural and social sciences, arts and humanities – it has extended well beyond the boundaries of the traditional observatory. Our new friends from Pyongyang could not attend the regular volcanological conferences, or access the subject's journals. Most tellingly, perhaps, they had not *seen* a volcano erupt. Multisensory experience of different volcanoes doing different things is what made Perret and Tazieff great volcanologists. I have enjoyed the opportunities to travel freely, to access the latest knowledge through the internet, and the academic currency of unfettered exchange of ideas and knowledge with colleagues around the world. I greatly admired Dr Kim and Dr Ri's achievements with such limited resources, but I also scratched my head sometimes, for instance when they asked me about drilling into the magma chamber to relieve pressure on it and forestall eruptions. This is science fiction for me – one of several topics of conversation that made me realise how science can suffer in seclusion.[25]

Meanwhile, James had been with the Korean seismologists setting up the earthquake sensors every six miles or so in a line outwards from Janggun Peak. There were six, each the size of a large paint tin and linked to a small GPS antenna and a data logger, all powered by a car battery and solar panel. I tagged along to watch one of the installations in a village called Sinmusong near the foot of the volcano. A small underground vault had been built for it in a potato field by a farmhouse, whose steeply pitched roof testified to the expectations of winter snowfall. James checked the instrument was working by stamping the ground, activated data capture and closed the vault lid. He wasn't interested in any rumblings of Paektu itself, but rather in picking up the vibrations transmitted through the Earth from large worldwide earthquakes. Seismic waves alter slightly if they pass through regions containing liquid, so James hoped to locate any molten rock beneath the volcano. But the sensors had to be in the ground for two years to register enough events to fill in the picture – a long period of nail-biting for James. His main worry was that the farmers might divert the seismometer's power to watch TV during power cuts.

*

I hadn't foreseen that when we would return to collect the sensors two years later, we'd be in the company of a sound engineer, a cinematographer and one of the world's most invincible filmmakers, Werner Herzog. In the intervening time, I had started working with Werner on a documentary film, *Into the Inferno*, an exploration of the nature and culture of volcanoes.

Though we were spoilt for choice with locations around the globe, we didn't begin production with the low-hanging fruit;

instead our first location was the DPRK. No Western film crew had ever been permitted to film on Mount Paektu, so this was a big deal for our colleagues at PIINTEC. Ms Ryu and her colleague, Mr Ri, worked miracles behind the scenes to organise film permits, transportation and interviews, and above all to win over concerned parties anxious about our intentions. I wanted to tell the story of the Millennium Eruption, and allow people to see James and the Korean seismologists at work, but above all I hoped to reveal how a volcano can embody the spirit of a people like a reverberation of the heart. Meanwhile, Werner's most ardent wish was to film schoolchildren singing in praise of the Paektu bloodline.

We had an even larger entourage in the Bae Gae Bong Hotel this time. As well as our colleagues and friends from the geological community and PIINTEC, there were two ecologists, a historian from the DPRK's Academy of Social Sciences, a representative of the Nature Conservation Federation, and the country's best interpreter, the kindly Han Myong Il. We all gathered in a meeting room, where I explained our hopes for filming. Werner then spoke of growing up playing amidst ruins in Germany after the Second World War. He talked about how, when Germany was divided, it became his passion to see reunification. When political efforts to this end failed, he had walked along the borderline 'to hold the country together'. Five years later, the Berlin Wall fell. Werner concluded by saying: 'My deep wish is that your country will be reunited. I don't know when it will happen, but I know it will.' It was a stirring homily and did much to win the confidence of our Korean friends.[26]

Clearing the treeline, on the way up the mountain with the film crew, I pointed out the summit. In an echo of Francis Younghusband, Werner exclaimed: 'This is not a volcano! It looks like somewhere in the Vosges mountains.'

'Wait until you see the crater,' I offered.

Twenty minutes later, we reached the lookout on the crater rim; even under a grey sky, he saw in all its hues and asperities the tumult of revolution. We filmed first with James and his Korean counterpart at Janggun Peak. James was delighted to find the seismometer still running after two winters. As we were set to move on, we heard a chorus of patriotic singing, and a marching procession of thirty or so young men in khaki uniforms emerged from the mist, their leader carrying a large red flag. They were not soldiers but students on a pilgrimage to the sacred mountain. As we filmed, they waved their red notebooks, threw their peaked caps into the wind, and sang heartily:

Let us go to Mount Paektu in spring and in winter!
Mount Paektu is the native home of my heart.

Back in Pyongyang, we filmed at a kindergarten, where all was purposefully choreographed. The visit culminated with a troupe of children on stage, singing: 'Mount Paektu is calling us, we will go there . . . even in dreams . . . to victory, until the end, following the Party!' Werner left the DPRK a happy man.

James was satisfied, too. With the seismometers recovered and the data backed up, he had a sensational haul. Several of the Korean geophysicists visited us in the UK to process and interpret the recordings, which proved exciting. First, they revealed a thickening of the Earth's crust under the volcano, testament to the millions of years of magma refuelling. More importantly, they identified a wide region containing molten rock about four miles down. This is the magma that might feed Paektu's next eruption.[27]

I was focused on a different problem – trying to tie down when, exactly, the Millennium Eruption happened. It puzzled me there was no historical date for it since there is no shortage of religious and bureaucratic records from Manchuria and

Korea dating to the medieval period. Why hadn't the zealous scribes of the era taken note of such a monumental event? Some historians had even suggested the eruption might have had something to do with the collapse in 926 CE of the Bohai kingdom, which spanned a vast area around Paektu until it fell to the Khitan armies of the Liao dynasty.

In fact, many attempts had been made to date the eruption from radiocarbon measurements of the trees entombed in the pumice deposits on the mountain. But these were imprecise, spanning at least the tenth and eleventh centuries. Then, I came across a remarkable study led by Japanese specialist on solar activity, Fusa Miyake, that made me think it might be possible to pinpoint the date to within a matter of months.[28]

Radiocarbon dating uses the decay of radioactive carbon as a timer. Radiocarbon is made in the atmosphere as a result of the Earth's continuous bombardment by cosmic rays from the sun and stars, and is taken up by living plants and animals. When an organism dies, its radiocarbon clock starts ticking down with a half-life of 5,730 years (meaning it takes that long for half of the radioactive carbon in the organism to decay). If you are starting to feel an inner glow, don't worry – only one in a trillion carbon atoms is radioactive, so you're safe.

You might think that all we need to do to work out the age of a biological sample is take a measurement of its remaining radiocarbon. That would be true, except that the flux of cosmic rays changes over time, so the production of radiocarbon in the atmosphere is not constant. This variation needs to be accounted for, and introduces a good deal of uncertainty in the dates. What Miyake and her colleagues did was to analyse the radiocarbon, ring-by-ring, in a 1,859-year-old cedar tree from Japan. The rings could be exactly dated to the year simply by counting back in time – one ring, one year – starting from the year the living tree was sampled.

As foreseen, the radiocarbon did not decrease straightfor-wardly back in time, but the team uncovered an exceptional shift in amount between 774 and 775 CE. The only explanation was that a massive flare from the sun briefly bathed our planet in X rays, generating a pulse of radiocarbon in the atmosphere. A fraction of this excess ended up in the living cells of the tree. Subsequent observations revealed the signature of the event in tree rings from around the world, pinning the timing of the event down to the Northern Hemisphere summer of 774 CE.[29] Such an intense episode of 'space weather' today would threaten much of our electronics infrastructure.

Why is all this relevant to the Millennium Eruption? Because some of the trees killed by the eruption would have been alive in 774 CE, in which case they could not fail to register the solar spasm. All I had to do was get hold of an old-enough tree from Paektu's pumice deposits and locate the radiocarbon offset – this would then anchor the year the tree died.

I eventually obtained the sample I needed – a larch tree, 264 years old when it died beneath a hail of Millennium Eruption pumice – and sent it to a specialist facility in Zurich. We hit the jackpot with our first round of measurements, pinpointing the ring formed in 774 CE. Then it was simply a question of count-ing the succeeding rings as far as the bark. In this way, we could say definitively that the last ring, which was complete, had formed in the year 946 CE. That meant the tree was killed between the summers of 946 and 947 CE. This was a huge step forward, but I thought we might do better still, as fallout from the Millennium Eruption had been reported in ice cores from Greenland. While their absolute dating is questionable, they record a very consistent seasonal pattern in sea-salt concentra-tion, related to winter storms in the North Atlantic. This showed that the volcanic dust fell in early winter, narrowing the erup-tion window to late 946 CE.

We could already reject the idea that the eruption triggered the fall of the Bohai kingdom – that had happened twenty years earlier. But I wasn't done. The trail now led back to Hiroshi Machida, who had been the first to draw attention to the Millennium Eruption. He had noted an entry in the chronicles of a renowned Buddhist temple, Kōfukuji, near Osaka in Japan. Dated 3 November 946 CE, it records that 'white ash fell gently like snow'. The scribes make no mention of any of Japan's volcanoes erupting at this time, and the temple is not far from Lake Suigetsu, where ash fallout from the Millennium Eruption was recently identified in sediments. Further, the Millennium Eruption ash *is* white – it is in the name Paektu ('white-headed'), referring to the deposits at the top of the mountain. Taking in all the evidence, and allowing twenty-four hours for ash clouds to reach Japan, I'll stick my neck out and state that the Millennium Eruption took place on 2 November 946 CE.[30]

With a firm date, there is now one record that stands out in the history of the Koryŏ dynasty. It says that in the year 946 CE, at the palace in Kaesŏng (280 miles southwest of Paektu, and a stone's throw from the Demilitarised Zone), 'the sky rumbled and cried out, and there was an amnesty'. We know from more recent eruptions, such as those of Tambora in 1815, that volcanic explosions are heard over huge distances. I think this is what created such alarm in the Koryŏ capital, leading King Chŏngjong to pardon prisoners. He is also said to have established twelve granaries, perhaps a state response to a food crisis after blankets of ash ruined winter crops.

Why isn't there more concrete information in contemporary Korean or Manchurian texts? One explanation, surely, is Mount Paektu's remoteness from the centres of power. Even in the late nineteenth century, at the time of Francis Younghusband's travels, it was hard to reach, surrounded by peatlands, criss-crossed by swollen rivers, snowed-in during the long winter, and its thick forests inhabited by man-eating tigers. Younghusband struggled to obtain food supplies, sunk to his knees in bogs that were impassable for pack animals, ached from traversing the endless succession of ridges, and was bedevilled by murderous midges, mosquitoes and gadflies. Apart from rare encounters with ginseng farmers and hunters, who knew the mountain's moods, he 'never saw anything but the trunks of trees'.

But another reason for the chronicles' silence on the Millennium Eruption could be that no eyewitnesses survived to tell the story, leaving only rocks, ice sheets and sentient trees to speak on their behalf.

★

Our collaboration with North Korean scientists continues, though all plans went on hold when the Coronavirus pandemic hit. With funding from the U.S. Government, James and our colleagues in Pyongyang plan an even more ambitious deployment of seismometers that will form a bigger network, with sensors installed across the border in China. This, James hopes, will help to answer a remaining question – why is Mount Paektu there in the first place? The volcano sits far from the nearest tectonic plate boundary, which runs east of the Japanese islands. It might have something to do with subduction, but that is far from clear. Meanwhile, I hope to explore some of the other volcanoes around Paektu, about which very little is known.

I imagine some will raise an eyebrow at the thought of any collaboration with the DPRK. We go to Mount Paektu because there are scientific questions to tackle, some of them with a direct bearing on volcanic hazards and humanitarian matters. We go there with the goodwill and direct support of many international partners in governmental agencies, learned societies and NGOs – in the United States, United Kingdom and China, as well as both North and South Korea. We go there because we have built a unique collaboration that shows it is possible to work across very different political systems and do good science. There are other kinds of international partnership and exchange involving the DPRK, but there is no comparable project working at our level, at least in geoscience. We go there because, despite the communication issues and other challenges that make collaboration anything but straightforward, we have built friendships and trust with colleagues who have a great thirst for knowledge. Everything we do stands or falls on the quality and impact of the research. We listen to each other.

The historian Ruth Rogaski has described Mount Paektu as 'sentient', and there is no question its history – and future – are entangled with territorial claims, foundation myths and geomancy. The volcano's meanings are multiple and malleable. But we should take a step back: these are all human constructs and projections – a stele thrust into the peak, imaginary dragons, unfeasible sexual unions ... Kosima, who has visited Pyongyang more than fifty times over three decades, once said to me:

> Human entities have no impact on the volcano – if we can elevate our timelines and our quest and recognise science as something that helps understand the environment and ourselves, then politics just becomes a factor in the mystery story of our own short lives. It is not a challenge for a volcano to sit on a border or to be part of a divided country! It belongs to no one. We're all only guests there.

She is right, of course: the physical, geographical masterpiece of the volcano stands aloof from all the attention it gets. But ... I return to that picture of two joyously beaming men, framed as if being raised on a plinth from the crater into an immaculate sky. This is not feigned emotion – this reunion was miraculous, the actualisation of a deep desire. I recall the historian who accompanied us to Janggun Peak saying: 'The dream of all Korean people will be realised when they can climb to this spot on Mount Paektu.'

There is such a thing as a geographical destiny of a country. One day, surely, peaceful reunification will come to the Korean peninsula. An utterly indifferent – but emblematic – volcano might just be what it takes to show the way.

Lava Floods and Hurtling Flames

The great geyser at Hekla, Iceland.[1]

'You cannot count how many volcanoes there are in Iceland – some are the size of frogs.'

– Werner Herzog[2]

Iceland is nothing short of a volcanologist's paradise. You cannot find a stone in the country that wasn't expressed through a volcanic orifice. The country *is* a gigantic volcano. One or another of the island's innumerable craters erupts every few years: over 200 events are catalogued for the period since the Vikings (and a few Celts) settled there around the year 870 CE, and the true figure may be closer to 300. With so much opportunity for igneous encounters, it is unsurprising that human fantasy and frailty are deeply engrained with experience of volcanic vistas and convulsions.

The island straddles Earth's longest mountain belt, the Mid-Atlantic Ridge, which spans the dark ocean between the polar regions. The ridge is forged from a partnership of volcanism and tectonics that continuously renews the Earth's crust.[3] In the North Atlantic, it divides the Eurasian and North American plates, which move apart two centimetres per year, creating a rent that is filled in by eruptions of lava along the sea floor. Worldwide, the new seabed created every year by this process of 'sea-floor spreading' adds up to an area about the size of Central Park in New York City. The Earth isn't expanding, so a comparable amount of ocean plate is consumed at subduction zones like the one offshore from Chile.

Beneath Iceland, the volcanism is supercharged by the melt expressed from a hot current of solid rock, known as a mantle plume, that rises slowly through the Earth.[4] Instead of lying one mile below the waves, as does much of the ocean ridge, Iceland's peaks reach as far above sea level. The result is an exposed tectonic plate boundary revealing parallel and cross-cutting fault escarpments, lava plains and countless volcanoes.

Icelandic volcanism achieved global notoriety in 2010 with the explosions of Eyjafjallajökull in the country's south. The intermingling of red-hot magma and glacial meltwater generated uncommonly violent blasts that shattered lava into powder. The ensuing clouds of floury dust headed for European airspace. Ever since a British Airways jumbo jet lost power in all four engines as it flew through an ash cloud over Java in 1982, the International Civil Aviation Organization has been braced for similar encounters.[5] It established nine Volcanic Ash Advisory Centres, hosted by national weather services around the world to monitor and advise on the threat.[6] All the same, the Eyjafjallajökull eruption caused global disruption to flights with economic losses counted in the billions of dollars.[7]

While this 'ashpocalypse' raised awareness of the airborne menace of explosive eruptions, we should take note of a more insidious peril of Icelandic volcanoes: lava floods. They armour the cone of a volcano but then ooze into lowlands, ravaging vast areas of farmland and pasture, infilling the nooks and crannies of a landscape like a gigantic road resurfacing project gone horribly wrong.[8] While explosive eruptions can begin abruptly and end in hours, giving little time to react, lava flows and floods tend to be more staggered and prolonged affairs, impacting first one locale, then another. This has elicited a variety of invocations and confrontations intended to stem or divert lava, or even to stop it bursting out of the Earth in the first place. The outcomes, though, have not always desirable.

Lava flows seldom kill people directly – they don't move at the speeds of *nuées ardentes*, and you'd have to be incredibly stubborn or unlucky for one to outrun you.[9] But they can pump phenomenal quantities of gases into the air, which can scorch and poison a much wider area with acid rain and fluorine (a little of this in your toothpaste is beneficial, but it is extremely toxic at high levels). This can despoil productive land, and it is in this way that lava eruptions have claimed far more lives through famine and disease than any explosive eruption in Iceland. More than that, the sulphur emissions from the largest Icelandic lava floods in history have cooled climate and triggered food shortages across the Northern Hemisphere in much the same way that Mount Tambora did in 1816, despite being very different kinds of eruption.[10]

The best understood example of a lava flood in Iceland is the Laki eruption. Its noxious emissions of gas and ash and the great tongues of lava that flowed into the lowlands devastated the food economy of southern Iceland. A fifth of the country's population perished from malnourishment and disease during what Icelanders refer to as the 'haze famine'.[11] The eruption persisted for months, and the haze became a 'universal fog' that spread over Europe and beyond in the summer of 1783. The statesman and polymath, Benjamin Franklin, attributed the severe European winter that followed to the airborne dust, while his contemporary, Erasmus Darwin (Charles's grandfather), thought it responsible for an 'epidemic cough'.[12]

Iceland's volcanoes were known to the outside world long before Eyjafjallajökull and Laki exported their effluvium. For centuries, Europe had been importing sulphur, mined from the country's geothermal spots. It was initially sought out for medicinal purposes, but its use in the manufacture of gunpowder made it increasingly popular.[13] Oddly enough, those dispatched by the firearms that made use of this explosive would, according to medieval theologians, have ended up in Iceland – in death that is – within one of its most active volcanoes, Hekla. Situated north of Eyjafjallajökull, it was regarded as the authentic location of purgatory.[14]

When Hekla itself reawakened on 2 September 1845, after a seventy-seven-year slumber, news spread to Europe with the ash clouds feathering the sky. In the Orkney islands, north of Scotland, a correspondent wrote of the people's astonishment at 'a great fall of dust'. A nuisance for those who'd hung out the laundry, but terrifying for the fishermen, who refused to go to sea the next day.[15] Near the volcano, the ashfall was so thick and the sky so black that farm workers lost their way home. That night, the mountain was lit with pillars of fire, and glowing lava spilled down its west flank from a string of apertures on the summit crest. Lava overran the water supplies of several farms, while more ash and clinker fell from the billowing plume, turning good pasture into a reeking mass of grit, acid and bleached grass. To make matters worse, the volcanic fallout carried so much fluorine that thousands of farm animals were poisoned.[16]

Since Iceland was a Danish dependency, the Hekla eruption caught the attention of leaders in Denmark. Acceding to the throne in 1839, King Christian VIII may qualify as the most ardent rock-hound of a ruler ever. He had a superb collection of minerals, knew Vesuvius well from his visits with the scientist Humphry Davy, and hosted scientific conferences. So it is less of a surprise that he assembled a small team of experts to investigate the Icelandic affair – this may even count as the first international scientific response to a volcanic eruption. The head of the expedition was the German geologist Sartorius von Waltershausen, who had recently finished a lengthy study of Mount Etna in Sicily yielding the first detailed geological map of a volcano.[17] The most famed scientist in the party, though, was his compatriot and celebrated chemist, Robert Bunsen. While better known for his research on gases and spectroscopy, he, too, had a passion for rocks.[18]

A fact-finding mission suited Bunsen well. He was an almost militant empiricist with a contempt for hypothesising, and was well prepared to undertake volcano fieldwork, since a gas chemist's profession then was beset with occupational hazards. He had only just escaped death by cyanide poisoning a year before while measuring effluent from a blast furnace in Derbyshire, England.[19] For Bunsen, volcanoes were just chemical engineering at a grander scale.

Before leaving for Iceland on a naval brig, Bunsen prepared a mobile field laboratory, taking care to ruggedise all his apparatus for rough travel and hostile conditions.[20] Arriving in Reykjavik, the first problem was to find good horses, since many were malnourished owing to burial of meadowlands beneath Hekla's lava and ash. But after two weeks, the party of scientists, guides and horses set off, crossing 'frightful lava flows, piled up in the wildest shapes, sometimes in clumps as high as a house, sometimes in masses . . . like cold, stiff dough, covered with countless wrinkles, and shot through with many fissures and hollows'. They first visited the mud volcanoes and sulphur mines of Krýsuvík before continuing to the northern side of Hekla, where they camped for ten days. Braving dust storms, treacherous cliffs and a thick, poisonous fog, Bunsen twice scrambled over cooling lava flows to the top of the glacier at the volcano's summit, the ice steaming under a blanket of ash. From there, he clambered down into the newly excavated crater to sample and experiment on fumaroles amidst heaps of fresh ejecta.

Bunsen's findings from his mission are impressive, including an explanation for geysers, a forerunner of the classification of lavas by silica content that remains the standard today, and recognition that some volcanic rocks are a blend of different magmas, like two differently coloured paints stirred together.[21]

The lavas Bunsen encountered were basalts – their very high temperatures make them runny like honey or tar. However the flow top cools quickly and becomes as sluggish as peanut butter, forming rucks and folds at the surface as the movement contin-ues.[22] This forms a crust that insulates the lava beneath, allow-ing it to overrun a much wider area than it might otherwise.[23] When the last gasp of magma drains away, tunnels remain.

Such tubes, or 'pyroducts' (your challenge: try to use that word in conversation today), offer unique underground experi-ences. They can be accessed where roof sections have collapsed, creating cavities. It was through one such chink that I climbed down into Iceland's most spectacular and intriguing lava cave system, Surtshellir. On this subterranean adventure, I was in the company of farmer and local historian Árni Stefánsson[24] and Michael Dunn, founder of 4th Planet Logistics, a company looking to facilitate human settlement of the solar system. According to the company website, their mission is to 'create and test habitats from naturally occurring terrestrial lava tubes analogous to the Moon and Mars in order to allow humans to find a new home in the future'. Michael told me he'd travelled to all but one of the countries of the world. Hard to trump, but it turned out I'd been to the only one not on his list: the Democratic People's Republic of Korea.

Surtshellir lies within the Hallmundarhraun plain in western Iceland. It was around the turn of the tenth century when the torrents of lava spewed from the foot of a vast volcanic table mountain stretching to the Langjökull ice cap, setting the stage for the Vikings' first serious magmatic encounter. A wide tract of good pasturage and fertile lowlands was obliterated. Five farms in the area recorded in *Landnámabók*, Iceland's medieval equivalent of the Domesday Book, are never mentioned again. Most likely they were swallowed up by the pitiless lava. If so, it is unlikely they will ever be unearthed from their basalt sepul-chre.[25] Today, the hummocky plain is brightened with miniature clumps of crimson-leaved blueberries, and dressed with the same pale mint-green plush of moss that covers much of the lava terrain of Iceland. Like a memory foam, it invites you to lie down and leave your impression on the landscape.

The lava tunnel entrance was hidden from view until we were more or less on top of it – its roof here was an arch of thick sheets of basalt. As we scrambled down a jumble of prismatic blocks into the gloomy vault, we found ourselves within a subway-sized passage. Michael explained his interest in the site: 'Lava tubes have all kinds of advantages for an extra-terrestrial home: you've got a ready-made primary structural shell that protects from cosmic radiation and meteorite strikes.' I was wondering how you would make them airtight, but he was on to that, and explained how habitable enclosures could be made by sealing the caves with inflatable membranes. He was evidently expecting *Homo sapiens* to become cavemen again.[26] He continued: 'We've got a great collaboration here with Árni; we'll test design concepts in this tube, just like the Apollo astronauts came to Iceland to experience something like lunar geology before leaving Earth.'[27]

Shining a torch around the underground cavity, Árni added, 'My great uncle, who lived on the nearby farm, found this extension to Surtshellir a century ago. The cave was decorated with beautiful lava stalactites – "lava straws", we call them. But sadly, nearly everything has been destroyed – look.' He directed his light at the ceiling, revealing a stucco of remnant stumps of lava 'icicles'. 'It was like gold fever – people took the straws and dripstones as souvenirs. That's why we use this cave for the Mars project – it's too late to protect it.'

Just as visitors in the modern era, whether tourist, volcanologist or space coloniser, are drawn to this cave, so too were the Vikings, more than a thousand years ago. But what prompted them seems to have much more to do with the underworld giant they saw as responsible for the eruption that formed it. The archaeologist who found their traces, Kevin Smith, reckons the cataclysm 'must have scared the bejesus out of people', and they wanted to make sure nothing like it ever happened again.[28]

Smith had known of Surtshellir for years.[29] A boat-shaped drystone enclosure of Viking style, said to have been an outlaws' hideout, had been discovered deep inside the tunnel. But what prompted him to survey the site urgently were reports that tourists were taking animal bones from a pile in the cave. In 2001, he salvaged the leftovers with Icelandic colleagues, and soon confirmed their Viking age. It was what he didn't find that led him to question the site's accepted interpretation. If rustlers or bandits had lived in the tunnel, they would have needed heat, light, water, some kind of dormitory, a latrine; all of which should have left residues: ash, soot, wax, bedding, nightsoil. The team returned in 2013, and this time recovered a veneer of soil on the cave floor. They found none of the evidence they would have expected had the caves been occupied; instead, the signs spoke of the occult.

Firstly, there were abundant chips of jasper used as fire-starters. Their chemical composition could be traced to distant Icelandic sources of the rock: the visitors to the cave had travelled from afar. Other finds were sourced from even more distant parts of the Viking realm – glass beads from Norway, and flecks of orpiment (an arsenic ore) from the borderlands of Iran and Turkey.[30] These were prestige materials that would have belonged to the pillars, not pillagers, of Icelandic society.

Radiocarbon dates show that the lava tunnel was entered as soon as it had cooled sufficiently, around 920 CE, and that it was used up to the first decade or two after Iceland's official conversion to Christianity in 999/1000 CE. This coincides strikingly with the year 1016, when pagan rituals were banned by law. In the centre of the boat structure, Smith and his team found a hearth with charred bones. On top of the fire ash were four small lead objects, including a cross. Since lead melts in a fire, they must have been placed after the last embers had extinguished.

Smith was now sure the cave wasn't an outlaws' refuge. For starters, Icelandic brigands and desperados had short life expectancies – they were usually hunted down in a matter of years. The new evidence was clear – the cave had been in use for a century before someone placed a Christian symbol on the fire pit. Further clues were found in the style of cut marks on the bones, and evidence for butchery in early spring (rather than the usual timing in autumn). These suggested sacrificial slaughter.

Then Smith realised the answer had been staring him in the face all along: the cave's name, Surtshellir. Surt's cave. Surt was a primeval giant who smote the goddess of fertility and agriculture, Frey, with a sword of fire before engulfing the world in flames. In another medieval source, Surt commanded hypogean fire creatures to rouse volcanoes. Significantly, he is only referred to in Icelandic literature: he is a unique character born of the Viking colonists' volcanic experiences, not a Germanic or Scandinavian import.[31] The earliest reference to Surtshellir, found in *Landnámabók*, relates a visit by a chieftain's son, Thorvaldur 'the hollow-throat', who travelled across Iceland's desert interior to chant a poem dedicated to Surt.

It seems clear now that the reason the Vikings entered the cave was to propitiate Surt, the personification of the lava floods that had engulfed the lowlands, and the murderer of Frey on whom harvests and prosperity depended. The animals were killed and dismembered at the cave entrance, and the bones were then carried to the liminal space by the hearth, where they were cremated to placate Surt and curb his destructive impulses. Generations of gatekeepers and officiators kept the ritual alive until paganism was forbidden.[32] Strikingly, no residues or artefacts from the cave date between the late eleventh and late seventeenth centuries, when the first explorers entered. It is as if people intentionally forgot Surt's address.

★

As if further to affirm their connection to the lively earth, Iceland's colonists established their parliament and court, the *Althing*, right on the tectonic plate boundary at Thingvellir (which translates as 'assembly fields'). Chieftains from across the country first gathered there in June 930 CE, a practice that continued in various guises until 1798. Sentences were handed out from the Law Rock at the foot of a prominent escarpment in the basaltic rock. Close by were pools where adulterers were drowned. Today, it is tourists that flock to Thingvellir National Park, titillated by the notion of planting one foot in America and the other in Europe (tectonically speaking).

Within a decade of establishing the *Althing*, an even more dramatic lava flood than that which moulded Surt's cave occurred in southern Iceland at Eldgjá (the Fire Gorge). Since the rituals at Surtshellir were likely underway by then, it seems that Surt had either found the burned offerings of mutilated sheep deficient, or he had got bored and relocated. Fed from a magma reservoir beneath Katla volcano (Eyjafjallajökull's larger, icier neighbour), basaltic lava spurted from multiple craters along a canyon-like fissure fifty miles long. It then spread out, like molasses, across the land, covering an area half the size of London.[33] As well as the immense carpet of clinkery lava, an even wider region was blanketed in frothy cinders that fell from ash clouds. Today, the landscape still leaks: water spills from the hillsides and cliffs through numberless clefts and spouts.

There are no direct historical references to the eruption. One of the first settlers in the area, Ásbjǫrn Reyrketilsson, was said to have dedicated part of the territory to the god Thor, naming it

'Thor's wood'. This suggests it was once a fertile, forested area, but in later medieval texts the area is described as a 'wasteland', implying a lasting transformation of the landscape. The best estimate for the year of the eruption was 934 CE, based on the approximate timescale of Greenland ice cores that contained fallout from Eldgjá. But the imprecise date made it tricky to identify any worldwide impacts. Then, by a stroke of luck, I managed to date it precisely.[34] Though I never expected it, the key was fixing the date of another eruption, of a volcano halfway around the world; once again I found myself face to face with Mount Paektu's Millennium Eruption. Or at least the record of its fallout in a Greenland ice core, which assigned the year 946 CE to that layer. Along with sporadic volcanic fallout, the ice cores also preserve a seasonal pattern of sea spray carried inland by winter storms. By counting how many rises and falls in sea salt there were between the Millennium Eruption and Eldgjá signals, I could tell the Icelandic eruption had fired up by the spring of 939 CE. What's more, the sulphur trace in the ice indicated the eruption may have persisted up to the fall of 940 CE. Some of the earliest settlers on Iceland, brought there as children, probably lived to witness the Fire Gorge aflame.

The destruction of prime upland grazing and lowland agriculture by lava and ash must have been disastrous enough. But greater distress and harm would have resulted from a more pernicious emission: the gases released into the atmosphere from the magma, including prodigious amounts of sulphur dioxide.[35] Acid rain likely corrupted a wide area of productive land, killing livestock and leading to widespread hunger and disease. Survivors would have fled the afflicted region in desperate hope of finding untainted food. While direct evidence of this is lacking in Iceland, further afield there are numerous entries in medieval chronicles that tally with the expected long-range impacts of Eldgjá's sulphur emissions. With the earlier date of

934 CE, none of these indications could be attributed to the eruption, but with the new firm timing of 939 CE, cause and effect snapped into place.

One of the most striking hints was a testimony in the *Chronicon Scotorum*, an Irish monastic chronicle of events and affairs from the earliest times up to the year 1135. Instead of the usual report-age – a battle here, a plague or the death of a king there – the entry for 939 CE begins: 'The sun was of the colour of blood from the beginning of one day to the middle of the following day.' Similar sightings of ruddy sunlight were recorded at an abbey near Rome, and another east of Paderborn in Germany. Almost certainly, these widespread accounts point to the lofty passage of Eldgjá's sulphurous dust veil. Such celestial marvels, including sightings of fireballs in the sky, had biblical connota-tions. They were regarded as dire portents, and faithfully recorded. A dark moon seen in eclipse is another sign of volcanic dust in the stratosphere. It tickles me to think that the monks who kept these annals could never have imagined how valuable their observations would be for those of us studying the impacts of volcanoes on climate and society a thousand years later.[36]

I wondered next if the dust might have chilled the globe. I turned to my tree-ring colleagues for help here, as there was no such thing as a medieval thermometer. The idea that tree-rings can tell us about past climate goes back to the 1940s, but the science of 'dendroclimatology' has developed tremendously over the past decade. The widths of individual rings, as well as their density variations and cell structure, can reveal much about growing season temperatures and rainfall. Quantitative records are built with reference to modern times, when we can measure the rings and compare them with reliable climate observations from satellites, weather stations and models. Importantly, tree-rings can be dated precisely to the year; so tracing your finger 200 rings from the edge of a tree sampled today takes you precisely two centuries back in time. The oldest living tree is a bristlecone pine in the White Mountains along the California–Nevada border. It germinated two centuries before the Great Pyramid of Giza was built. Dead trees that have been preserved in peat bogs, or simply exposed on the ground for centuries or even millennia, can also be fitted into a precise chronological framework, pushing back some tree-ring records to the end of the last Ice Age.[37]

When we looked at the tree-ring data for the mid-tenth century CE, we found a clear signal of stunted growth across Central Europe, Scandinavia, the Canadian Rockies, Alaska and Central Asia for the year 940 CE, during the Eldgjá eruption. It added up to one of the coldest Northern Hemisphere summers of the last 2,000 years. Turning back to the chronicles, there was indeed plenty of evidence for a time of hunger and famine across the world. For instance, accounts from Sicily tell of a food crisis and abandonments of fortresses and countryside, while a Lombard historian mentions devastating food shortages in Italy. Further accounts of starvation are found in sources from France, Switzerland, Germany, Iraq, the Maghreb and China. Collectively, many thousands of lives may have been lost, and though human factors and failures usually go a long way towards explaining famine, the sulphur unbound from the Earth by Eldgjá likely had a hand, too, in these calamities.

But what still seemed odd to me was that in Iceland itself, it appeared that nothing was written of such an exceptional episode – the effects had to have been felt over much of the country. I then came across a text in the *Poetic Edda*, also known as the *Codex Regius*, Iceland's most important medieval manuscript. The first poem in the collection, *Vǫluspá*, sees the god Odin wake an ancient seeress from the dead. She recounts the history of the world before foretelling the fate of the gods and giants. When I read the following lines portraying *Ragnarök*, the Nordic version of the end times, my jaw dropped:

Comes Surt from the South with the singer-of-twigs,

. . .

'Neath sea the land sinketh, the sun dimmeth,
from the heavens fall the fair bright stars
gusheth forth steam and gutting fire,
to the very heaven soar the hurtling flames.

. . .

will grow swart the sunshine in summers thereafter,
the weather, woe-bringing: do ye wit more, or how?[38]

Could they describe volcanic fire fountains blasting skywards
while palls of ash obscure the sun and heavenly bodies? We
already know what Surt can do – and 'singer-of-twigs', though
it sounds rather genial, refers to 'fire'. The link between volcanic
dust veils and cool summers is well understood, so the reference to woeful weather is also striking. Do these lines connect
volcanoes and climate eight centuries before Benjamin
Franklin's perceptive remarks? The poem also tells of 'drops of
poison' coming through roofs, which is reminiscent of acid rain
falling from a volcanic cloud. Everything here speaks to me,
with almost eyewitness clarity, of the apocalyptic scale of the
Eldgjá eruption – this verse is not describing run-of-the-mill
volcanic activity.

The seeress ends by explaining that a 'great godhead' comes
down to govern all after the dust settles. This is highly sugges-
tive of the Christianisation already underway in Iceland at the
time of the eruption. Is it possible that the poem set out purposefully to rekindle harrowing memories of a volcanic haze famine
to loosen the hold of paganism? If so, a volcano may have helped
convert a people to Christianity.

★

Volcanoes don't observe lockdowns – on the contrary, they continued to unlock their potency throughout the Covid-19 pandemic. This meant that many of my colleagues in observatories had to figure out how to continue monitoring and hazard assessment from home. The episode that had me transfixed during the UK's third national lockdown was an eruption near Reykjavik. It began on 19 March 2021 in the southern Reykjanes peninsula, where the volcanoes had slept since the end of the last Ice Age, around 12,000 years ago.[39] Even the northern half of the peninsula had not been resurfaced by lava since 1240. Thanks to live internet streaming, I tuned in frequently to follow the creation of the new volcano, named Fagradalsfjall. In the first week, magma boiled and spattered into a snowy sky from an elfin black chimney while vividly sanguine lava oozed downslope. So petite was the tableau, I imagined relocating it to my garden at home (though it might have menaced next door's gazebo).

To my mind, this eruption offered a unique opportunity to monitor gas emissions that would be the closest you could feasibly get to a mid-ocean-ridge volcano. It is extremely challenging to measure volcanic fluids on the deep sea floor, for understandable reasons, but here was a proxy on land, less than an hour's drive from the Icelandic capital. Seeing people in their hundreds gathered at the spectacle – among them chefs sizzling hot dogs on hot lava, troubadours, and partners taking marriage vows – it was evidently easy to reach the site (often half the battle of fieldwork). Over the next three months, I worked with the University of Iceland and the Icelandic Meteorological Office to support gas-monitoring efforts while still at home in the UK. They sent me photos and data from Fagradalsfjall so I could work out the gas composition. I could almost smell the sulphur!

Of course, doing this in my loft office was a poor substitute for being there but I was still thrilled to feel part of the action again after such a long hiatus. When Iceland was designated a 'green list' country for travel, I took my chance. Even as a volcanologist, it is not every day you get to see lava flooding the landscape, and I was itching to see an eruption like Eldgjá's, albeit a scaled-down version.

I arrived on the summer solstice. The sun set after midnight and rose three hours later, though hidden behind a low cloud deck. At this time, there was no community transmission of Covid-19 in Iceland and there were very few restrictions. I was collected mid-morning by colleagues from the Icelandic Meteorological Office and University of Iceland. Climbing into their Land Rover, I couldn't help grinning and blabbering owing to the frisson of being in an enclosed space with people after such an extended period of isolation.

Our first undertaking was to measure the flux of sulphur dioxide coming from the volcano. The equipment was the same as that my group had developed twenty years earlier at Masaya volcano in Nicaragua – an ultraviolet spectrometer pointing up at the sky and recording continuously, and a GPS unit tracking position as we drove along a quiet road skirting the shore of the pretty lake Kleifarvatn, searching for the plume. When my nose caught the first whiff of sulphur I blurted out, 'That must be it!' One of the volcanologists, Melissa, corrected me: 'No, that's the fumaroles at Krýsuvík, just over there.' She pointed to clouds of steam emanating from the foot of a nearby ridge. I remembered this was where Robert Bunsen had measured volcanic gases 175 years earlier. He and von Waltershausen had camped beside a bellowing steam vent that served as both bath and kitchen range, as well as yielding valuable samples.[40] Once again, I had the strong sense of following in the footsteps of volcanology's trailblazers.

After several runs back and forth beneath the plume and alongside the lake, we continued to the new eruption site not far away. A rough track led to a crest overlooking the wilderness of newborn lava. Stepping out of the Land Rover, I was beaten back by a bitter and drizzly wind coming straight off the North Atlantic. It was clear at once I was underdressed, and I whined that the forecast had looked favourable. 'This *is* good weather,' my local accomplices replied.

We couldn't get close to the erupting vent owing to the great apron of fresh lava spread around it, but we found a line of sight to the glowing magma, which was spilling from a breach in the cone to feed a zigzagging lava flow. The moving lava disappeared into hidden pools and tubes, insulating it and enabling it to engulf all the low ground as far as I could see. This was an uninhabited area, but I could imagine the economic catastrophe if instead farmland and pasture were being consumed beneath the burning rock.

The surface was mottled and textured with blooms of sulphur and other minerals, and brick-red patches of *'a'ā* lava in a grey sea of the smoother variety known as *pāhoehoe*. Both sorts get their names from Hawai'i. They are easy to distinguish and present different challenges for the hiker – crossing a solidified *pāhoehoe* lava is a bit like walking on a greenhouse roof, while the loose clinker of *'a'ā* is perilously unstable. How such different fabrics arise from the same molten stuff remains only partly understood.

We were aiming to measure gases with an infrared spectrometer, just as I had on Montserrat. Once the equipment was set up, one of the geochemists, Andri, said, 'Watch the lava in the crater.' After a few minutes, I noticed the level rising and magma surging into the channel. 'It has been doing this for a month and a half,' he continued, 'but for the first weeks it was much more spectacular with a pulsing lava fountain. It was so high, we could see it from Reykjavik.' Indeed, I had seen photographs taken from the capital with the cathedral's white steeple looming in the foreground and a similarly shaped red spire of lava shooting above distant hills. It drove home the proximity of the volcano to the capital city, and oddly juxtaposed heavenly and infernal emblems. Andri added: 'What's more remarkable is the gas composition changes rhythmically with the fountaining episodes. There may be a cavity under the cone that periodically fills with bubbly magma, empties like a geyser and refills. It's as if the volcano is breathing.'[41]

After measuring for an hour, we packed up and drove down a saddle in the ridge until the track expired in a wall of fresh lava. 'This should warm you up – it looks like you need it,' Andri said, grinning. 'We used to drive this way to get near the crater, but now it is impossible'. I jumped out and walked right up to bulbous lobes of silver-grey basalt that had recently congealed against a stony incline. The lava was about chest-high and mostly solidified, but I still felt the fierce radiance from fiery gleams within fractures in the rocky bulk. A chorus of clinks rang out as razor-thin shards of volcanic glass splintered off the flow's cooling and wrinkling rind, here resembling tree bark, there a slop of entrails. The distant slopes and peaks shape-shifted in heat shimmer above the fresh igneous mass.

We returned to collect another set of data a few days later, joined this time by Gro Pedersen, a geologist leading the effort to survey the expanding tract of lava. Her work is critical to identifying where the flow might head next. Gro and her team were mainly working with aerial photographs, but she also needed to take survey points with GPS on the ground.[42] Now and then a gas monitor clipped to her backpack beeped: 'That's carbon monoxide coming off the soil and vegetation cooking under the lava,' she explained. 'It's odourless but poisonous, so I have to carry this alarm.' As we skirted the edge of the flow, walking upslope, Gro summarised how the lava had initially been confined in one valley, but had now poured into several others in the corrugated terrain. Lately, the output of lava had almost doubled to ten cubic metres per second, which would fill four average houses in a minute.

After half an hour, we rounded the crest over which the lava had spilled into the valley, and came face to face with a battle – not quite *Ragnarök*, but an echo of it perhaps. The operators of an excavator and bulldozer were pushing up earthen barriers to hem in the lava and prevent it from spouting over the other side of the ridge, against which it was lapping. Gro pointed behind us towards the sea: 'If it overflows here, then the road to Grindavík will be cut.' Looking at how much terrain the lava had already swallowed, it seemed pretty clear who would win if the eruption went on much longer. It turned out later that the lava supply cut off before the ridge was overtopped. But prior eruptions on the Reykjanes peninsula had lasted, in fits and starts, for centuries, so no one thought it was all over. Sure enough, the fire giant, Surt, renewed the blaze the next year.

*

Obstructing lava is one way to counter a volcano; oblations are another, but more offensive interventions have also been practised. In 1669, Mount Etna was threatening the city of Catania on the Sicilian coast, with lava flows issuing from a vent low down on the flank of the volcano. Several villages had already been destroyed when some fifty Catanese citizens – equipped with picks and axes, and shielding themselves from the furious heat with water-soaked hides – hacked open the wall of a lava channel to redirect the flow. They succeeded, but then ten times as many angry townsfolk from nearby Paternò showed up, armed to the teeth and demanding the breach be stemmed: the lava was now headed towards *their* neighbourhood. The lava channel healed itself, though, and the basaltic slag resumed its inexorable passage to the sea. It partly surmounted Catania's city walls but the damage within them was relatively contained. Afterwards, it was declared that anyone who tried to divert lava would be liable for unintended damages, an edict later written into law.[43]

Three centuries later, in 1973, a new volcano, Eldfell, was spewing lava torrents towards the only town on the Westman island of Heimaey, off the south-west coast of Iceland. It threatened to ruin the harbour, and most of the population was evacuated to the mainland as cinders buried their homes.[44] Earth barriers were quickly constructed to blockade the lava, but then more direct action was taken to extinguish it. With the help of firefighters and volunteers, seawater was pumped onto the advancing flows. The idea was to stiffen the molten rock, making it pile up rather than spread out. In the end, the harbour was spared – and, if anything, improved – thanks to a tongue of lava that exuded along the seabed and formed a breakwater. Moreover, when the refugees returned, they extracted the geothermal energy in the fresh rock to heat their properties. But volcanologists were divided on the influence of the water cooling – some thought the eruption had just stalled of its own accord.[45]

If sprinkling water on a lava flow is wasted effort, and turning up with agricultural tools will only inflame the neighbours, what's next? Incredibly, even aerial bombardment has been tried as a means to stop an eruption in its tracks. In 1911, Frank Perret was writing routine bulletins on Kīlauea's activity in Hawai'i. They were published in the newspaper of an influential Honolulu businessman, Lorrin Thurston, who helped to fund the volcano observatory. Thurston also wanted to make money out of a hotel at the crater's edge, and took a close interest in countering the threat of lava flows. He proposed detonating explosives lowered through apertures in lava tubes to stem the current, forcing lava to the surface. If this were done high up the volcano, he argued, then the flow would start over on a new path, buying time downhill.

In late December 1935, an eruption of Mauna Loa on the Big Island of Hawai'i had been underway for a month when it took a turn towards the city of Hilo. Thanks to a well-developed lava tube, the front of the flow was advancing so fast it looked as if it would reach Hilo within two weeks. Thomas Jaggar, the director of the Hawaiian Volcano Observatory, wondered if it was time to try out Thurston's plan. He put in a request to the U.S. Army to bomb the lava flow from the air. The next day, a telegram arrived from Franklin Roosevelt that read: 'Planes authorized. Good luck . . .'

Ten Keystone biplane bombers arrived, along with observing aircraft; targets were selected, and 600-pound bombs were loaded. Jaggar watched from a safe distance as they were dropped in pairs. Five direct hits were recorded, squirting glowing lava into the air like 'a geyser of blood'. One bomber's wings were pierced by the liquid shrapnel. The flow came to a stop the following day, and Jaggar felt vindicated in calling in the military, but in truth the eruption had already run out of puff.

In his last book, which was published posthumously, Jaggar speaks often of the notorious Caribbean volcano, Pelée, but oddly overlooks Pele, the sometimes vengeful goddess who animates the Hawaiian volcanoes. Before Europeans arrived on the Big Island, a priesthood and temples were dedicated to her veneration and propitiation. For centuries, it seems, people lived in harmony with her caprices. The bombing of the lava shocked many in the native Hawaiian community – it violated the sanctity of Pele's realm. Some people were not surprised when, less than a month later, two of the Keystones touched wings and burst into flames while flying low over the airfield in Pearl

Harbour. Six men died; the two who miraculously survived were the only crew members who had not been on the Mauna Loa mission.[46]

★

It was just one day after Fagradalsfjall stopped erupting on 18 September 2021, that magma reached the surface on La Palma in the Canary Islands. Fifty years had passed since the touristic island had last witnessed an eruption and the new episode gave just a week's warning. Whereas Fagradalsfjall served as a festival venue and tourist attraction, onlookers at La Palma's Tajogaite eruption witnessed entire communities – homes, shops, churches, schools, roads, everything – consumed by lava. By the end of the affair, nearly 3,000 buildings were obliterated.

A maelstrom of human activity mirrored the volcano's turbulence. Dozens of local, national and international organisations – civil defence, police, fire brigades, scientists, refugee agencies, journalists – flocked to the scene to monitor, manage and report on the crisis. I was among this mêlée soon after the eruption's onset, with a small team there to sample lavas, and despite my years studying volcanoes arrived underprepared for the scale of unfolding ruination and human suffering.[47]

Only on Montserrat have I witnessed such destruction of the built environment. But there it was the result of pyroclastic flows and mudflows in a crisis that spanned years. Here on La Palma, everything happened in a matter weeks – some residents got out with minutes to spare as their homes were engulfed in red-hot slag, from foundations to roofs. This overpowering spectacle became yet more apocalyptic beneath black plumes that developed as acres of plastic sheeting covering banana farms went up in flames. Meanwhile, evacuees were further affronted by wide circulation of drone footage of their smouldering properties.

As the eruption dragged on, the faithful rallied for a Mass then joined a procession led by a local priest bearing the statue of the island's patron saint. Prayers were offered for the volcano to calm. Though the eruption continued for another two months, and despite the immense dislocation and disruption, only one resident is known to have died. He fell to his death while cleaning ash from the roof of a building.

La Palma is a very different place to Iceland but seeing people lose every material thing that had meaning to them made me think more deeply of the trauma of those dispossessed by the lava floods of Hallmundarhraun, Eldgjá, Laki and Hekla, or by similar eruptions of Etna, Kīlauea, and many other volcanoes. In the tally of endeavours to put out the fires of the Earth, it is unclear how effective any of the methods are, whether intercessions or interventions. I'd tended to think that trying to stem or deflect such a merciless geological force is futile but now I see why people are prepared to put up a fight against the volcano.

Red Sea, Black Gold

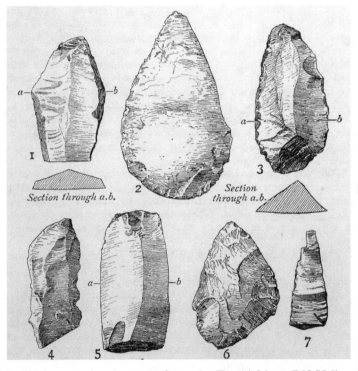

Neolithic stone implements from the East African Rift Valley.[1]

'. . . nature run amok also makes culture, makes artifacts.'
    – Susan Sontag, *The Volcano Lover*[2]

As I watched flamingos skimming the ruddy water of Lake Langano against the hazy backdrop of the rift escarpment, a jet-black object in the sand, about the size of a razor blade, caught my eye. I picked it up and recognised it as a flake of obsidian. There was plenty of the glassy rock to be found in lava outcrops on nearby Aluto volcano. However, this piece had been deliberately fashioned into a tool. One side had beautiful ripples from a single conchoidal fracture; the other displayed several prismatic facets running lengthways. These caught the sunlight as I turned the piece round.

To discover and hold a stone artefact is at once thrilling and evocative. It sparks a mystical connection between your hand and that of its manufacturer and user across thousands, even millions, of years. Stone tools are far more durable than wood or shell or hide; they are not predated on like flesh and bone. It is rocks that survive to tell us what humans were doing and even thinking a long, long time ago – where they lived, what they ate, how they interacted, and how they solved problems.[3] The blade I had found in the Ethiopian Rift Valley was probably made by a Neolithic pastoralist around 4,000 years ago.

Obsidian comes from eruptions of viscous lavas that lose all their gas and congeal without crystallising – it lacks bubbles and minerals. A chunk bashed off an outcrop splinters like glass because it *is* glass, just a natural rather than manufactured variety. Bash a bit more carefully, and you have an edge far sharper than a medical scalpel, suitable for slicing most foodstuffs. Some archaeologists refer to obsidian as 'black gold' owing to its prestige in prehistoric times. It is among the resources that volcanoes offer up.

While archaic humans already exploited obsidian a million years ago, it was our own species that learned to fashion it with flair in Africa in the Middle and Later Stone Age over the last quarter of a million years. The very word 'obsidian' has an African connection, as we learn from Pliny's *Natural History*: it derives from Obsidius, a Roman consul who is said to have discovered the rock in Ethiopia.[4] Figuring out how to turn brittle obsidian into precision implements may even have helped to drive wider social, cultural and behavioural change. I wonder if our Stone Age ancestors loved it for its lustre and gleam as much as its keen edge.

Obsidian is extremely useful, too, for modern-day archaeologists, who use it to reveal the human geography of the unwritten past. A magma reservoir in the Earth's crust evolves chemically and physically over time. It cools and slowly crystallises. Denser minerals settle out, depleting the remaining molten rock in magnesium and iron, while enriching it in silicon. The magma can also melt the rock at its walls, and sometimes two different magmas connect and mingle. These myriad processes impart a particular bouquet to the molten rock – like wine, magma is defined by its *terroir*. The equivalent of a sommelier for magma is an igneous geochemist. A skilled one could immediately discriminate between a full-bodied Californian andesite and a fruity Auvergne mugearite.

However, most volcanic rocks contain assemblages of common minerals, such as olivine, quartz and feldspar. If handed a piece of lava and its bulk chemical composition, our connoisseur would struggle to identify the individual volcano that gave birth to it. Crystal-free obsidian, on the other hand, has a unique geochemical fingerprint defined by the amounts of different elements. If two obsidian tools have the same print, they were made with the same source rock; if a tool's print matches that of an outcrop, that is where the raw material came from.

The technique was pioneered in the 1960s for the Mediterranean, where rich obsidian sources are found on Sardinia, Lipari and Melos. For an archaeologist finding obsidian tools in Tunisia, for example, it is as if they are stamped 'Made in Italy'. Obsidian sources are found in most of the volcanic belts of the world – including the Americas, Oceania, the near East, and Northeast and Southeast Asia – so the method is widely applicable. It provides direct evidence for the ranges over which people could exchange ideas as well as pieces of stone, yielding unique insights into prehistoric human culture, mobility and cooperation, even in the absence of a single human fossil.[5] For a volcanologist, it offers a perfect point of collaboration with archaeologists.

And there is more to volcanoes and human prehistory than obsidian. As the archaeologist Mary Leakey wrote in her autobiography, Eastern Africa is a place 'where nearly every exposure [of rock reveals] some archaeological or geological excitement'.[6] The Rift Valley is a palimpsest of the sediments of volcanoes, wind, lakes and rivers along with faunal relics and traces – it is a time machine that can take us back to the world of humans before humanity. Picture, for instance, three short figures walking together across a blanket of gritty brown ash recently blasted from a vent on the flanks of Ngorongoro volcano. This is the image conjured up thanks to Leakey and her team unearthing their tracks at Laetoli in Tanzania, some 3.7 million years later.[7] Their age was impressive enough but most astonishing was the lack of any indentations from knucklebones – whoever left these traces walked upright. The footprints were attributed to *Australopithecus afarensis*, the closest known relative of our genus, and named after the Afar region of Ethiopia, where the bones of 'Lucy', the emblem of her species, were discovered.[8]

But how do we date this prehistoric vignette in the first place? The answer is in potassium-rich crystals contained within the ash. A tiny fraction of the potassium is radioactive and decays gradually into argon gas – the half-life of the process (the time taken for half the potassium present to decay) is more than a billion years.[9] As soon as a volcanic rock is erupted and cooled, the argon generated starts building up in the crystal. By measuring how much is there, the age of the eruption can be calculated. This is an extremely powerful way to piece together the deep volcanic history of a region, bearing in mind that radiocarbon dating can only reach a few tens of thousands of years back.[10] Just as importantly, it indicates the age of archaeological materials, including stone tools and human fossils, which are not readily dated themselves. We know Lucy's antiquity not from her skeleton or teeth, but from the estimated ages of ash fallout found above and below her level in the sediments.[11]

Sometimes ash deposits far from their volcanic source are too fine-grained to contain any crystals. There is still a way to date them, though, using the same method of chemical fingerprinting applied to obsidian. Most of such a fine-grained ash deposit will consist of tiny pieces of volcanic glass, whose composition can be measured and compared with ash deposits nearer the volcano that are coarse enough to contain datable crystals. If a chemical match is found, then the far-flung ash has the same age. I was part of the team that used this approach to show that the oldest uncontested *Homo sapiens* fossils, which come from southwestern Ethiopia, are 30,000 years older than once thought.[12]

What emerges from these studies is that individual volcanoes have long lifespans – often hundreds of thousands of years, most of which they pass by slowly crumbling apart through the action of earthquakes and weather. *Homo sapiens* only emerged as a species within the last 300,000 years, which means that where people have long lived in the shadows of volcanoes, the volcanoes were in most cases there first. They inspired and enabled our ways of life, and sometimes brought devastation. Our stories are intertwined.

★

The Rift Valley of Eastern Africa is one of Earth's greatest geological features. It spans the tropics from Mozambique to the Red Sea – the distance between Los Angeles and New York – and ranges in height from the glaciated crater of Kilimanjaro to the searing lava plains of Afar. The Rift's shoulders funnel rainfall to vast freshwater lakes. This manifold geographic template, shaped by tectonic and volcanic forces over the past forty million years, enabled ecosystems to accommodate the climate teleconnections of successive ice ages, leaving us today with one of the most biodiverse regions on Earth. Some scholars even think the variegated landscape and its sporadic rejuvenation by eruptions and earthquakes helped to stimulate human evolution by providing abundant fresh water, rock shelters, and natural lava barriers for corralling prey.[13] Others look at the Rift Valley and Red Sea coastline as natural migration corridors that enabled our ancestors to expand beyond Africa around 70,000 years ago, either through the Levant or Southern Arabia.[14]

One of the first to grasp that this topographic rent resulted from 'rifting', or stretching, of the Earth's crust was the geologist John Walter Gregory, who explored Kenya in 1893. He integrated geological findings with botanical, zoological and ethnographic observations in his masterpiece, *The Great Rift Valley*, published three years later. In recognition of his pioneering work, he had one of the world's largest topographic features named after him: the Gregory Rift, which stretches 700 miles from northern Tanzania to northern Kenya. [15]

But the rifting does not end there – to the north it cuts through Ethiopia's high plateau then becomes more diffuse in the scarified volcanic and hyper-arid lowlands of the Danakil Desert of Afar. Here it bifurcates, one branch connecting to the Red Sea through Eritrea, the other to the Gulf of Aden through Djibouti. Studies of this region in the 1960s helped to propel plate tectonics theory into mainstream thinking.[16] Why this tectonic mother lode developed right here is still not fully understood, but several factors seem to have conspired, including pre-existing fault lines that buckled due to the force of hot rock upwelling from the deep Earth. This so-called 'mantle plume' has been at work for a long time: thirty million years ago, it sourced the incalculable volumes of basalt lava now exposed in the dissected highlands of both Ethiopia and Yemen.[17]

Although geologists recognised the importance of Afar for understanding the birth of oceans from the incision of continents, it was not until 1971 that the first comprehensive scientific survey of the area was undertaken, co-led by none other than volcanologist Haroun Tazieff.[18] With a combination of helicopter support, camel caravans and legwork, and the backing of Afar's sultan, his team obtained the first picture of the region's volcanism and tectonics. A highlight of the expedition was the discovery of bubbling lava lakes within the crater of a volcano called Erta 'Ale.[19] But just as the studies were gathering pace, the toppling of Emperor Haile Selassie in 1974 and the installation of a barbarous military junta, the Derg, abruptly ended foreign access to the region.

My visit to Lake Langano during a six-week tour of Ethiopia came about thanks to a window of opportunity (also known as unemployment) in 1993. The Derg had been overthrown a couple of years earlier, so travel was becoming possible again. But my main inspiration had been a charismatic geologist, Seife Berhe, with whom I shared an office. His stories of growing up in Addis Ababa, the Ethiopian capital, and recollections of adventures with the Geological Survey of Ethiopia, thrilled me so much I had to see the place for myself. After my first visit, I soon wanted to go back in order to rekindle the long-stalled volcanological studies in the Danakil Desert in north-eastern Ethiopia, and returned the following year with funding for a small-scale expedition. Maybe it was too soon after the years of conflict; Eritrea, which shares the territory of the Danakil, had also just gained independence from Ethiopia, destabilising the border zone we were headed for. Maybe it was just bad luck. Whatever the reasons, this would prove to be a near-disastrous mission.

Things didn't start off too badly. Seife joined me in Addis Ababa, along with some of his former colleagues from the geological survey. We planned to collect rock samples, perhaps even of molten lava from Erta 'Ale's fiery lake, and to carry out a very precise GPS survey that, when repeated, would reveal the pace of Afar's tectonic stretching.[20] After several days in the capital obtaining permits, and stocking up on supplies, we set off. We stopped for the night in Metehara, at the foot of Fentale volcano, then tracked the course of the Awash river in the morning, passing close to where Lucy had been unearthed twenty years earlier. By late afternoon, we arrived in Asaita, the traditional seat of the Sultanate of Aussa, and straddling the Awash, where hippos and crocodiles played vanishing acts. The dauntless British explorer, Wilfred Thesiger, came this way as a young man in 1933, three years after attending Haile Selassie's coronation. The Danakil Desert was still, even then, little known to Europeans, and Thesiger was seeking the terminus of the Awash river, which puzzled foreigners since there was no sign of it reaching the Red Sea.[21] We called by the regional government office, where we secured the services of two guides, Mahé and Mohammed, and obtained typed letters of introduction for each of the five local administration zones through which we would pass.

Vultures eyed us from telegraph poles as we climbed into the cars at sunrise the next day, while doves idling in the road shied as they sensed our approach, then flew beside us like porpoises chasing a ship. Before long, we were off the main road and following a rough trail between long basalt bluffs, from which rust-coloured boulders had spalled. The rocks, fault scarps and volcanic landforms were so exposed and vivid it felt like crossing a three-dimensional geological map. We saw no one, but passed many rock tombs of the Afar, built from slabs of lava and circled with stones. According to Thesiger, a couple of pillars at the top indicated an avenged death. In his day, the Afar were renowned for castrating their enemies and keeping the 'trophies'. Seife asked the guides over our camp dinner if the custom prevailed. Mahé shook his head, but one of Seife's colleagues, with more precognition than we realised at the time, suddenly exclaimed that we must be crazy to undertake such a journey.

By lunchtime the following day, we reached the Danakil Depression, the harshest part of the desert. Ever since leaving Addis Ababa on the Ethiopian plateau, we had been gradually descending, and we were now below sea level. It was ten degrees hotter, up to 40°C (104°F), and the sky had bleached with the thickening air, from azure in Addis to bright powder blue. The vehicles waltzed over fault-cleaved basalt lava fields until the trail emerged on to a playa encircling Lake Afdera. Even though I was expecting it, the sight of such a vast body of water in one of Earth's fiercest deserts took me aback. From a distance, the water looked milky, bleached by haze, but it turned aquamarine and then turquoise as we neared it. By mid-afternoon, after resolving an encounter with quicksand, we had reached the shore and found a shady spot for our camp.

In the morning, an elderly Afar man appeared out of nowhere, carrying a plastic teapot and listening to a radio slung around his shoulder. He introduced himself as Ibrahim, the chief of the area. We showed our paperwork and explained our business and he agreed to help us, though he made clear he would not accompany us to Erta 'Ale on account of its *jinn* – evil spirits. We wanted to scout out a route to the volcano and headed north across the plain, passing a wide cone whose truncated top spoke of wave action. It had formed when the sea filled the Depression, which it did until around 30,000 years ago, when magma intrusions and eruptions of Alid volcano to the north formed a barrier to the Red Sea.

After an hour, we reached lava flow lobes emerging from windblown sands, their ropy surface burnished by centuries of desert exposure. These marked the southernmost extent of the volcanic range. I felt like I was exploring anatomised basalt tracts of the seabed on a planet where a baking sun had evaporated all the oceans – which was, essentially, what had happened right here after the Danakil was cut off from the Red Sea. Beneath the plain lies an immensity of salt, the residue of the sea's desiccation. It amazed me, then, when Ibrahim showed us a freshwater spring where the Afar bring their goats. Our water supplies were limited, so we did not hesitate to slake our ever-present thirst, despite the carpet of goat poo around it.

The GPS measurements involved finding suitable open-air sites where we would bang survey nails into the rock and record their position with pinpoint precision. While setting up the equipment on an ancient lava flow, two figures shivered into view in the glittering air and approached us across the plain. The two men, when they reached us, claimed to be representatives of the Afar Revolutionary Democratic Unity Front. Both carried AK-47s on their shoulders and wore traditional curved daggers in their belts. One of them, Habib, fidgeted with a hand grenade, while the other cleaned his teeth with a *miswak* twig. They were agitated and demanded to speak with Ibrahim. A prolonged and heated negotiation followed. I recalled an account by Werner Munzinger, the first European to cross the Danakil in 1867, a journey all the more remarkable for taking place in the suffocating heat of summer. He wrote of his encounters with tribesmen:

> Scientific ends are not understood . . . They think we should never encounter all these dangers without some strong motive, such as treasure to be found, &c., so that we cannot be astonished when they make us pay heavily. [22]

No one comes from so far away for nothing, and we agreed a fee with Habib. Nevertheless, I felt uneasy, recalling that Munzinger, his wife and a small army were later massacred by the Sultan's men on another crossing of the Danakil.

The following day, just as we were preparing to go to work, Habib and his companion appeared unexpectedly in our camp. Along with Ibrahim and our guides, they formed a circle in the shade of the palms. Each sat on his haunches, carbines and Kalashnikovs to hand. What followed I gleaned from translations via Seife and body language – it boiled down to an attempted kidnapping. At one point, Habib suddenly jumped up, shouting and brandishing his gun. Mahé, who was also armed, immediately faced him off. The others urgently gestured to both to sit down, and the passion passed.

The argument continued after it got dark, until Ibrahim announced that we should abandon our mission and head back in the morning. We readily agreed. I crawled into my sleeping bag anxious we might be strafed with machine-gun fire in our sleep, but the high suspense had been exhausting and, after ten minutes of wild imaginings (swimming across the lake to safety as bullets flew overhead), I slept surprisingly soundly.

We turned tail after a hurried breakfast – any feelings of dejection from cutting our mission so short were tempered by the sense that we had got out by the skin of our teeth. It was a shame, though – we had made the best arrangements possible for the mission, had all formal authorisations and the service of resolute chaperones. I had been concerned all along about Mahé's rifle – it hadn't seemed in the right spirit to go into a land I didn't know with a lethal weapon on display – but in retrospect, I believe it helped to avert a more unpleasant outcome. Seife thought we were lucky to be alive: 'Those men are used to killing,' he said. On the other hand, having read the accounts of Munzinger, Thesiger and others who had travelled in the Danakil over the past century and a half, our experience seemed typical to me. The Danakil's environment is extremely harsh, the resources few. When outsiders have come, it has often resulted in appropriation of land, water or minerals. Xenophobia is a natural adaptation – and it is true that we, too, were seeking something of value.

What would have happened in the distant past, I wondered, when survivors were fleeing eruptions that inundated whole sections of the Rift Valley with ignimbrite? Did they end up amongst communities with different customs? And, if so, could this have been one of the ways we honed our social networking skills, and innovated our material and spiritual culture?

Sadly, I didn't complete the GPS measurements. If the few survey nails are still in the ground, they should have shifted around a stride apart after thirty years of ongoing stretching of the crust.[23]

★

Despite the failure of my mission to Erta 'Ale, I was not put off studying the Danakil's volcanoes. For a while, though, I did it more safely, using Landsat images and declassified spy satellite photos, from which, in the comfort of my office in Cambridge, I compiled a timeline of Erta 'Ale's activity since the 1960s.[24] It was while pouring over these data that a volcano east of Erta 'Ale, near the Red Sea coast in Eritrea, caught my eye. There are innumerable volcanoes in the region, but most are armoured in dark basaltic lavas. This one, called Dubbi, had prominent lava flows, too, but they were superimposed on a bright terrain that looked like an apron of pumice or ash. Pretty soon, I found what I was looking for in the archive of the *Times of London*, a brief report from 1861 relayed by Captain Lambert Playfair, assistant Resident at the British colony of Aden (now part of Yemen):

> On the night of the 7th or morning of the 8th of May, the people of Edd were awakened by the shock of an earthquake . . . At sunrise . . . fine white dust fell over the village like rain . . . it increased to such an extent that . . . the whole place was covered with the dust nearly knee deep . . . At night we saw fire and dense smoke issuing from a mountain called Jebel Dubbeh . . . Nothing of the kind had ever been heard of before, and the people were exceedingly frightened.[25]

Thunderous detonations were heard for hours around the southern Red Sea, including in the Eritrean entrepôt of Massawa. Although the Suez Canal was not yet constructed, the seaway was already of great strategic and commercial significance, and contested by Ottoman, Arabian, Egyptian, Abyssinian and European powers. Given the prevailing atmosphere of international rivalry and intrigue, it's hardly surprising the explosions were widely misapprehended as artillery bombardments.[26] Ships were launched to confront the enemy but soon they were plunged into darkness by the ash cloud spreading above. The captain of one steamer described encountering a 'London fog' of white dust and being unable to see the length of the ship.

In a later report, Captain Playfair communicated the vivid account of two Somali brothers who climbed the volcano while it was still agitated. Guides led them to the foot of the mountain but were too frightened to go on. At the summit, the brothers found more than a dozen fuming and fiery craters. One pit blasted out rocks, while water spurted from other orifices like 'waves of the sea'. As I continued reading the report, I came to a passage that surprised me greatly:

> The names of the villages burned which were located near the mountain, are Moobda and Ramlo. One hundred and six men and women were killed and their bodies were not found. The number of animals killed is unknown.

The region has always been very sparsely populated and pumice falling from the sky is a discomforting experience but seldom lethal – how could so many have perished? Looking again at the satellite image, I was sure now that the pale ground was the fall-out from the 1861 eruption, but the lava flows were clearly on top of it, so must have occurred later. This was beginning to look like a very significant eruption that could even have resulted in much wider climatic impacts. I discovered that the worst recorded flooding in California took place in January 1862 – could that have been linked to Dubbi's stratospheric dust? But I couldn't go any further with historical sources. Nor could the space images tell me how thick the deposits were – essential information for any calculations of the size of the eruption or scale of its gas emissions. Without being able to map and measure the pyroclastic layers, there was no more that could be said of the nature of the explosions. I would have to go to the volcano itself to find these things out.

Captain Playfair's last letter published in *The Times* on the Dubbi eruption included a plea for more concerted investigation of the volcano 'to give to the scientific world an accurate description of this interesting phenomenon'. Getting on for 140 years later, in 1998, I did my best to fulfil that wish, in the company of Pierre Wiart, one of the first PhD students that I supervised. Our aims were to investigate the deposits on and around Dubbi to get a clearer picture of what had transpired in 1861, as well as to collect rocks that might help us understand the volcano's existence – with its record of explosive activity, it was clearly a different beast from most of the volcanoes in Afar, like Erta 'Ale, which produce lava floods. The journey would only sharpen my interest in the intersections of social and physical landscapes.

The capital of Eritrea, Asmara, is perched at the crest of the Ethiopian plateau. The country had been an Italian colony, and Asmara retains a fabulous legacy of 1930s Art Deco architecture. There is the FIAT Tagliero, perhaps the world's most important gas station, and the Cinema Impero, which was screening the 1965 epic *Battle of the Bulge* on my first visit, and whose bar still serves perfect espresso. With help from Eritrea's Ministry of Mines and Energy, and Seife's brother, we obtained the services of a driver, Mesfin, and a four-wheel-drive vehicle, as well as all the necessary documents authorising our passage as far as Asseb in the far south of the country. Despite the experiences of my last excursion to the Danakil, I didn't feel any anxiety to be returning – I was just excited to be heading into another area I'd become familiar with but only from the vantage point of a satellite's eye.

We left Asmara late, reaching Massawa well after midnight and checking into a hotel that was inexplicably overrun with frogs. By lunchtime the next day, we could see the Gulf of Zula and clusters of dark cinder cones. This is precisely where the

Red Sea's oceanic ridge propagates inland through the Danakil Depression to connect with the Erta 'Ale range. Sandwiched between this arm of the Rift and the southern Red Sea lie the Danakil Alps, a range of metamorphic rocks estranged in the tectonic break-up of Arabia and Africa. Nearby, at Abdur, we stopped to inspect an ancient limestone reef that now lay above sea level. It had formed around 125,000 years ago, and encased countless obsidian flakes and blades, along with disarticulated fossil oyster and crab shells – the trash from a Middle Stone Age beach party. [27]

The route now followed the mangrove-fringed Red Sea coast. The air sizzled, and the plastic jerrycans we'd bought in Asmara were proving to be a terrible false economy. They were deforming between the scorching metal struts of the roof rack, splitting at the seams. Water and fuel trickled down the windows, attracting so much dust we had to use the windscreen wipers. After repositioning the containers several times, everything reeked of diesel. By the time we stopped for the night at the coastal outpost of Gelalo, the vehicle was squalid inside and out; all our food boxes, cutlery and supplies were sullied.

Three years had passed since my brush with Habib and his comrade, but the memories were fresh. So my heart sank when, as we were packing up in the morning, I spotted someone approaching us on foot, a rifle across his shoulders. I braced myself for confrontation. But the gun turned out to be a shepherd's stick, and on reaching us its elderly bearer simply asked if we had broken down and needed help.

By day's end, we emerged from the other side of the Danakil Alps onto a powdery plain. We pulled up beside a lonely cinder cone and clambered to the top. In front of us stretched a volcanic desert fantasy in pastel shades from grey to pink, flecked with clumps of acacia trees that traced dry watercourses. Most imposing (and enticing) on the horizon was Nabro, the Danakil's largest and highest volcano, truncated by a sloping caldera rim. But Dubbi's appearance did not disappoint. To lay eyes for the first time on a volcano you have come to know well but only from space images is thrilling. There is the rush of instant familiarity, but only now do you perceive its aura – the nexus of terrain, sky, past paroxysms and social landscape. And we were not the first to survey the scene from this spot, judging by the abundance of obsidian tools caught in the volcanic gravel.

We made for the toe of one of the lava flows on Dubbi's west flank that I had marvelled at in the satellite data. It's thickness – about the height of a five-storey building – was impressive. We noted the details and started collecting rock samples. A bluff of pristine pyroclastic rocks at the foot of an eroded cinder cone also caught our eye. There was no doubt these were ash and pumice deposits of the 1861 eruption: the archives of the eruption we were looking for. A Chilean colleague once said to me that such an outcrop is like the black box flight recorder on a plane – the depths and angles of layers, the shape, roughness, orientation, dimensions and composition of constituent particles – all speak to the intensity, duration and magnitude of the eruption.

The deepest bed was the uppermost, and it was a meshwork of angular, walnut-sized white pumices. It was easy to dislodge pieces. This was surely the sediment from the most intense phase of day one of the explosive eruption, which had piled ash upon Edd and turned day to night. We eagerly set to work, measuring, sieving and weighing the deposits; taking notes and photographs; and bagging representative samples. Two Afar men came over, curious to know our purpose. One said his great grandfather had been killed by the 1861 eruption, adding that the mountain is still haunted by *jinn*.

Impressive as the thick fallout of pumice at the top of the section was, the underlying deposits were even more striking. They featured thinner beds of rounded pumices and laminated ash. The smoothness of the fragments indicated they had rubbed against each other, knocking off any edges, which meant this was the deposit of a flow and not another layer of fallout from the sky. We realised we were looking at the deposits of *nuées ardentes*, which at once explained the high mortality and lack of traces of victims during the 1861 eruption. *Nuées ardentes* sweep across the ground too rapidly for anyone to escape, and often bury what they engulf. Somewhere beneath the surface must lie the remnants of the villages of Moobda and Ramlo, and the bones of the herdsmen and women who dwelt there.

When we summited Dubbi itself, we found the Somalis' testimony from 1861 fitted with the view before us of an inverted backbone of vents and rents from which lava had poured. What surprised us more was our estimate of the lava volume, which easily ranks the eruption as the largest in the recorded history of Africa.[28]

I had one more day before flying out of Asseb (Pierre and Mesfin would stay on for another week). I'd felt the nearby volcano Nabro calling me the whole time – its presence was powerful, and I felt an urge to see what lay within the caldera and beyond. A little reconnaissance could do no harm, I thought, and would pave the way perhaps for future study. Judging the distance to the top, I felt confident there was time to take a peek and make it back before sundown. Pierre and I set out at first light, following a dry streambed cut into grey lava. Here and there, collected in indentations in the rock were small pools of water – always a miracle in the desert. Then we followed a spur leading to the peak of Nabro's wide crater rim. Now there were patches of open woodland and squat palm trees prospering in rocky niches – it seemed a different world to the scorching plain below.

Pierre was taking his time to inspect the rocks, and I was soon some way ahead. By around 10am, as I was approaching the brow of the caldera, a brilliant white cube perched on dazzling drifts of pumice came into view. I was taken aback and could only think of the monolith in *2001: A Space Odyssey*. I'd not seen any structures on the mountain in the satellite images, and was perplexed. Then I noticed an aerial, an encampment, and soon a dozen soldiers. They were just as startled to see me.

I broke into a grin. They smiled back, and one of the men brought me hot tea in an enamel mug. But the cheerful welcome quickly soured: why was I there? Why alone? Where were my papers? (They were in the car.) I explained the circumstances of my visit and expected Pierre to arrive at any moment. But he didn't – I would find out later that he had run out of puff and returned to camp. I asked the soldiers to contact the Department of Mines or send someone to locate Mesfin. Instead, they contacted the Ministry of Defence in Asseb. Worrying about Pierre, and impatient to move the situation on, I started walking back down. But two soldiers grabbed my arms and another raised his machine gun, shouting, 'You see what this is? When you come here, your programme is cut!'

I was furious and exasperated. I was also baffled as to why there should even be a military command post on a volcano in the middle of nowhere. Had they got the idea from the Bond film, *You Only Live Twice*? It was not the ideal moment to ask. Nor to appreciate the breathtaking view across the caldera as I was led away. In spite of the circumstances and the thick desert haze, I could make out the Erta 'Ale range and the peak of Ale Bagu volcano, and even the distant crest of the Ethiopian escarpment rising the other side of the Danakil Depression.

I was kept under guard in a rudimentary barrack room by two teenage conscripts for seven hours, all the while simmering with resentment, concerned about Pierre's whereabouts, and wondering if I would make the weekly flight from Asseb. Eventually, an officer arrived. He placed his revolver on a table, sat beside me, and asked about my mission. Shortly, he announced that I was free to go, but since it was late to walk down, I would be escorted by vehicle. I was flabbergasted that a road existed – it was hidden from view, carved into the sheer rock wall of the caldera. My new friend confided that he had always hoped to study in England, and asked if I might help him get to the British military academy, Sandhurst. It was an affable denouement to a frustrating episode. After a round of hand-shakes with my captors, I was led to a waiting Hilux. With the light fading, I made mental notes of the roadside geology as we descended into the crater by a series of switchbacks, aiming for a low point the other side. Crossing the rugged terrain in the centre of the caldera, I spied an alluring outcrop of burnished obsidian. I had to come back one day.

Just five months later, Eritrea and Ethiopia went to war following a border dispute – now I understood why there was a military post on the highest point along the frontier, and why a 1960s war movie featuring defensive lines, spies, tanks and infantry was being screened in Asmara. The conflict was cruel and profitless, and left a legacy of landmines and enduring enmities. It would take an entirely unexpected turn of events to present the possibility of returning to Eritrea.

★

Although we had barely sampled any rocks on Nabro, Pierre Wiart and I wrote a paper in which we speculated on the volcano's origins and nature of its lavas. It had to draw mostly on analysis of satellite images, and left me wondering what else we might have found had we been allowed more than a glimpse of Nabro. In a way, it mattered little – no one else was working in the area, so hardly anyone read our article. Until 12 June 2011, that is, when Nabro suddenly and violently exploded.[29]

One of the first scientists on the scene was Seife Berhe, who had relocated to Asmara. A colossal lava flow was snaking across the plain beyond the flanks of the volcano when he arrived, and dark ash clouds were mushrooming from the caldera. This event, the first-ever recorded eruption of Nabro, opened a window of opportunity to resume field research in the region, which had been impossible for foreigners since the border war. This time, I teamed up with three Eritrean geologists and the British seismologist, James Hammond – it had only been a few weeks since James and I had returned from North Korea. We obtained authorisation from the Eritrean government to make an urgent survey of the volcano and to install a network of seismic sensors to record ongoing rumbles. But in addition to our studies of the fresh eruption,[30] I found an opportunity to connect the present with the deep past of Nabro – the answer was to be found in the glittering obsidian I had spied under military escort.

We reached the site two months after the eruption ended and found an astounding scale of impact on the landscape. Lava flows had inundated a huge tract along the border with Ethiopia, engulfing many homesteads. Several new cones had formed, some still glowing at night, and a wide apron of black clinker had buried two villages inside the caldera, displacing thousands of people. Acacia trees that would have towered over me were almost completely buried – only the outermost twigs of their crowns poked above the grey cinders. Somehow they had survived, for new leaves were already unfolding.

As we sampled the freshly erupted rocks, I also kept my eye out for ancient obsidian. This was because I'd recently worked the other side of the Red Sea in Yemen, with the archaeologist Lamya Khalidi. She was studying Neolithic and Bronze Age sites in southern Arabia, and most of the tools uncovered there were obsidian. My assignment in Yemen had been to locate the volcanic sources. Since obsidian lavas are extremely viscous on eruption, they pile up into stubby lobes. This makes them easy to spot in satellite images, and I had readily identified several candidates on the Yemeni plateau. Sure enough, when we visited our targets on the ground, we discovered spectacular outcrops of obsidian, and even a prehistoric quarry where the raw material had been mined.[31]

From the chemical make-up of our Yemeni samples, measured back in the lab, we found clear matches with scores of tools from Neolithic and Bronze Age sites across the southern Arabian highlands – and, impressively, as far as 600 miles away in Oman.[32] That is a phenomenal range of human interaction, not least given the aridity of the region, and speaks to the high status of good-quality obsidian. But oddly, we found few matches with any of the implements that Lamya had collected from surveys and excavations along the Red Sea coast of Yemen. Even though they would have been just a few days' walk from our sources, lowland peoples had sought most of their obsidian from elsewhere.

With my fixation for the vitreous rock now verging on the obsessive, and better able to recognise an artefact thanks to working with Lamya, I couldn't help noticing stone tools and prismatic 'cores' (from which blades had been struck) almost everywhere on and around Nabro. In some places, a scatter of flakes on the ground suggested that someone had sat down and spent a while whittling the rock. It mystified me to think that this Stone Age refuse had lain at the surface undisturbed, perhaps for tens of thousands of years. I made a modest collection of the artefacts, and knocked fist-sized samples off the geological outcrops with my hammer, in an echo of the prehistoric miners who must have done much the same to obtain their raw material.

I sent the tools to Lamya. Though unimpressed with my less than systematic survey, she was still able to classify a range of scrapers, blades and cores, and suggest they were made by Neolithic people 4,000–5,000 years ago. Some, she explained, had been 'retouched' – this meant the margins had been carefully serrated to make them suitable for cutting or scraping wood, bone or hide. Comparing how shiny the edge was to the rest of the stone surface, this had evidently been done at a much later time, probably after the tool was worn out and had been discarded by its original owner. The green economy has deep roots.

Again, we had the tools as well as the geological obsidian samples fingerprinted in the lab. The analyses revealed three chemical groups amongst the archaeological samples, all of which corresponded to local sources on the volcano. This was no surprise – many of the implements had probably been made expediently on the spot. But to our astonishment, we also matched Nabro obsidian to nearly half the tools from the coastal sites in Yemen that Lamya had been puzzling over for years.[33] As the crow flies, these were 120 miles away – not so far compared to some of the other ranges we had uncovered, but no one swims across the Red Sea carrying hunks of obsidian, nor does the volcanic glass float. The only way it could make the journey is by some kind of boat. By linking the tools in Yemen to lava flows on Nabro we had demonstrated the existence of a seafaring people capable of transporting the 'black gold' from Africa to Arabia more than 4,000 years ago. This illustrates the power of obsidian provenancing – it is very unlikely a Neolithic or Bronze Age raft will ever be found at the bottom of the Red Sea, but the presence of 'Made in Eritrea' tools in coastal Arabia decisively reveals prehistoric maritime interaction.

More than that, Lamya believes that the availability of obsidian stimulated the growth of this network, helping a new culture – the Neolithic lifestyle of pastoralism and pot-making – roll out across the Horn of Africa. At the same time, African traditions of tool design and manufacture extended into Arabia – it was a two-way diffusion of knowledge, technology and culture.[34]

The trading opportunities of the Red Sea fought over by imperial powers in the time of Dubbi's eruption seem like a distant echo of the reordering of the prehistoric world through the interactions of distant peoples and the exchange of a sought-after volcanic commodity.

Had I looked harder, I think there would have been evidence for exploitation of Nabro's lustrous obsidian at much earlier times. I sent several of the pumice samples to a lab in New Mexico for argon dating. The results paint a picture of the key stages in the life history of the volcano, stretching back 400,000 years, before the emergence of *Homo sapiens*. The most workable obsidian lavas were erupted 160,000 years ago on the flank of the volcano. I cannot say if they were quarried then, but at around the same time, further south in the Kenyan Rift, people were already using and carrying obsidian over long distances.[35] Any extraction of Nabro's obsidian would have been curtailed for a long time, though, after the volcano exploded 130,000 years ago, forming the ample caldera and plastering the surrounding lowlands with ignimbrite.

The volcano had another convulsion 62,000 years ago. This was responsible for the banks of pumice at the summit where I had stumbled upon the military post. Soldiers posted there had helped with the emergency evacuation during the volcano's 2011 eruption, which began at around midnight after five hours of ground shaking. Of the more than 10,000 people who were displaced, most had been living in the villages within the caldera, whose elevation high above the roasting plains provided a cool sanctuary and attracted rain clouds. They had stored rainwater in cisterns covered with sticks, irrigated food crops, kept sheep and goats, and even soothed aches and pains by bathing in warm fumaroles. The Red Sea and its marine resources were a day's walk away. Now, their villages are a Pompeii of the Danakil; their footprints preserved under piles of scoriae. How many earlier human traces might lie within older layers of ash and lava? As I have so often found, risk and reward go hand in hand on volcanoes.

Given the antiquity of obsidian use in the region, and Nabro's commanding aspect, I feel sure the volcano has attracted attention since the first humans laid eyes on it – just as it lured me up to its sky-high crater. In the later prehistoric period, it anchored a regional economy and stimulated cultural innovation. Today, it is not obsidian that draws prospectors but the geothermal energy stored in superheated groundwaters circulating beneath the volcano.[36]

Working there – close to the sparkling (and actually turquoise-coloured) Red Sea, amidst both steaming lavas and ancient ignimbrites, meeting refugees who had fled their buried homes within the caldera, stumbling across so much hand-crafted obsidian, and encountering ageless rock tombs and the rusting residue of modern warfare – gave me a profound sense of the physical, social and even political prominence of the volcano. It had grown, collapsed, mutated, eroded, healed and regrown over hundreds of thousands of years, during which our own species evolved, perhaps even driving our evolution down particular paths. With its eruption in 2011, it had petrified history before my eyes.

But this was only the beginning of what Africa would reveal to me about our ancestral reliance on volcanic landscapes – it was in another scorching African desert that I found even more extraordinary evidence of volcanoes sustaining, and even nurturing, human life.

Water Tower of the Sahara

Trou au Natron, Tibesti mountains of Chad.[1]

*'In the desert, life's pleasures are simple but very real: a long
drink of clean water.'*
– Wilfred Thesiger, 'A camel journey to Tibesti'[2]

It's not every day that you meet an astronaut, especially one who turns out to have had a hand in your research. But it happened to me entirely unexpectedly in the early 2000s aboard a sailboat in the Gulf of Mexico. The encounter was with Captain Michael Baker, veteran of four space missions. I was excited to be able to tell him that I was studying experimental radar imagery recorded from the Space Shuttle. His reply took me aback: 'I commanded that mission and operated the radar. I collected those data!'

Radar devices both transmit and receive electromagnetic energy using an antenna. Think of air-traffic control radar, for example: a rotating aerial beams out microwave energy, and the illuminated display shows the location of planes that reflect some of the signal back. The strength of the 'echo' depends on the size and shape of the aircraft.[3] But radar does more than this – it is also sensitive to the target's 'roughness'. A smooth-topped ash deposit looks dark in a radar image because most of the microwaves are scattered away, while a bouldery lava flow reflects more of the beam back to the sensor and appears bright. NASA was interested in this capability for mapping the geology of the inner planets, notably the spectacular igneous terrain of Venus.[4]

The region I had been looking at via this radar imagery wasn't on Earth's sister planet, but it might as well have been, since it was barely easier to reach: the Tibesti massif of the eastern Sahara desert. Although one of the largest volcanic hotspots on our planet – ten times the area of the Big Island of Hawai'i, or about the size of Switzerland – the region remains barely known scientifically, owing to decades of conflict and a legacy of land-mines.[5] I was curious to understand how the Tibesti volcanoes, like Mount Paektu, had formed so far from a tectonic plate boundary, and this required understanding their geological history.[6] As I explained my interests to Captain Baker, he smiled and added, 'That range is the most striking land feature I ever saw from space – a huge, dark smudge in the largest hot desert on Earth, right in the heart of Africa.'

I had long felt a calling to desert travel, fuelled by reading books such as Wilfred Thesiger's *Arabian Sands*, and seeing movies like *The Flight of the Phoenix* and Werner Herzog's *Fata Morgana*. I'd even looked into taking a Land Rover maintenance course in my twenties, contemplating a trans-Sahara trip. But my particular fixation on the Tibesti arose from conversations with a retired colleague, Dick Grove, who had travelled there in 1957. Dick's reminiscences thrilled me; he spoke of travel by camel, ambitious volcano ascents, nights so quiet you could hear shooting stars, and stumbling across spectacular ancient rock art.[7] On top of that, as impressive as it can be to tour the Earth from a computer, you can't collect geological specimens that way. And Dick had none for me to work with: 'All my rock samples were lost – our army drivers tossed them out of the Land Rovers, for some reason.' I was desperate to follow in Dick's footsteps, to hammer the rocks, to see the volcanic strata, but the security situation on the ground only ever seemed to deteriorate, with continual violence and political unrest making the region near impossible to access.

I was astonished, then, in 2015 when I spotted the following report on the website of the scientific journal, *Nature*[8]:

Fears of terrorism and political instability put some scientists off joining geologist Stefan Kröpelin on his latest expedition to unlock the secrets of the Sahara desert. One of the most experienced Sahara researchers in the world, the 63-year-old geologist from the University of Cologne in Germany sets out on 19 February [2015] on a six-week trip to the remote Tibesti mountains in northern Chad ... But the mountains are also close to the Libyan border where terrorists are rumoured to be operating training camps.

I was agog. Not only was Kröpelin bound for the Tibesti, but according to the article he planned to reach its most emblematic volcano, Emi Koussi, the highest point in the Sahara and with a caldera colossal enough to enclose a city the size of Boston. It brought back memories of drooling over it in space images, which have often been my gateway drug to real-world adventure. I was so impressed by Kröpelin's plans that my instant reaction was to write an email wishing him success.

I wouldn't have to salivate much longer; the following year, I found myself substituting for a scientist who *had* been put off joining Kröpelin on a follow-up mission. And, this time, Kröpelin was keen to have a volcanologist on the team. Until now, I had only thought of the Tibesti as an unreachable geological wonder – the chance for me to go was the fulfilment of a fever dream. I would soon discover that the dark smudge in the Sahara has a human significance as profound as its geological roots.

★

The extent to which climate has shaped world history has become something of an academic quarrel. Many natural scientists studying past climatic fluctuations recognise the influence of environmental change on pre-industrial societies. On the other hand, most historians explain the rise and decline of states and empires in terms of political, economic, social and cultural factors. Both perspectives are valid, but anyone who denies that environment plays *any* role in human affairs needs only to compare maps of world population density and agricultural productivity: societies flourish where there is food and water. And there is very little of either of these in deserts.

While few people live in the Sahara, a region as vast as the U.S., tens of thousands of migrants traverse it annually, hoping to find a new life in Europe. Some do not make it – their vehicles break down in the sand. Without water, human survival can be counted in days. In 2017, more than forty men, women and children perished when the truck they were travelling in broke down in northern Niger.

But the Sahara has not always been so inhospitable. Today's desperate migrants follow prehistoric trails established when the Sahara was less arid. In the eastern Sahara, these converge in the Tibesti mountains, strategically connecting northern and central Africa and the Nile Valley. Dick Grove had been one of the first to recognise the geological evidence for a once-wetter Saharan climate in ancient lake and river sediments in the Tibesti – lakes and rivers only exist if there is rain or snow. But radiocarbon dating in the 1960s was still in its infancy, and it was unclear if these deposits were younger or older than the last Ice Age, which ended around 12,000 years ago.[9] Thanks to Stefan Kröpelin's research, it is now clear that the rocks Dick studied date to between 8500 and 5300 BCE.[10] This interval is now referred to as the 'African Humid Period' – or, more lyrically, 'the green Sahara' – and it came about through shifts in tropical rainfall belts, largely driven by a 26,000-year-cycle in the orientation of the Earth's axis that made for wetter summers.[11] But think not of the Sahara transformed into lush jungle; imagine instead an open savannah landscape like the Sahel today: good for hunting and grazing.

Today, beyond the banks of the Nile and a few scattered oases, the eastern Sahara is so devoid of moisture that it is almost completely unpopulated. The mere presence, then, of a stone tool or fragment of pottery in the sand bears witness to a once more favourable climate. The onset of the 'green Sahara' was fast-paced: dunes turned to scrubland over, at most, a few centuries, providing subsistence for hunter-gatherers. By 7000 BCE, nomads arrived, bringing sheep and goats. But things shifted back around 5000 BCE, when the desert reclaimed its territory. By 3500 BCE, the rains had all but stopped. Human survival demanded mobility – the pastoralists and foragers had to keep moving to follow the rains. Stefan Kröpelin thinks the resulting 'climate migration' stimulated the rise of civilisation along the Nile in Egypt and Sudan.[12]

But even in the hottest latitudes, a few ecological niches remained, most notably in the Tibesti, the Sahara's largest and loftiest massif, whose stature still brings rainfall to this day. Flora and fauna, including people, found refuge there when hyper-aridity gripped the region, as it has done repeatedly during past ice ages. When the Sahara greened they could extend their ranges again.

The inhabitants of the Tibesti are known as the Tubu. Their ancestry is not found in chronicles – it is inscribed in rockfaces. The rugged terrain of their mountain stronghold helped to forge their skills in raiding and defence.[13] It was the lure of a region 'all but unvisited and unknown to the outside world' that brought the British explorer, Wilfred Thesiger, to the Tibesti in 1938.[14] While it remains mysterious and inscrutable for outsiders even today, the Tibesti's importance as a trans-Saharan nexus saw the Tubu interact with whomsoever showed up over the centuries – Arabs, Ottomoans, Germans, French, Italians, British – as colonial, expansionist and religious spheres waxed and waned.

Being situated at the north-western frontier of Chad, the Tibesti lie closer to Libya, Niger, Algeria, Sudan, Egypt, Nigeria and Cameroon than to the levers of central governance in Chad's capital N'Djamena. Conflict has plagued the region since the country's independence from France in 1960. Following the Arab Spring and downfall of Muammar Gaddafi in Libya, the situation on the ground only became more unstable, with the rise of extremist groups in neighbouring countries. A gold rush in the Miski valley, which bisects the south-eastern Tibesti, added to tensions, with thousands of migrant miners skirmishing with the Tubu over scant water resources. The Tibesti is now a hub for gangs, rebels, mercenaries, kidnappers and traffickers; a conduit for weapons, drugs and people.[15]

Given this violent backdrop, the UK Foreign Office advised against all travel to northern Chad. This was a challenge for me in planning my trip to join Stefan Kröpelin's team, as I had to have a risk assessment for the expedition approved by my university. I emailed Stefan for help and was amused by his reply: 'I have never done a risk assessment in my life.' Somewhat reassuringly, he added: 'But you don't need to worry about terrorism except in the capital.' I formulated a suitable evaluation of threats: snake bites, heatstroke, landmines, kidnapping, vehicle breakdowns, and so on. It was formally approved, though it was made clear to me that my insurance policy would not cover hostage negotiations. After experiences like my brush with Habib and his friend in the Danakil, and aware of the rise in kidnappings elsewhere in the Sahara, there was, naturally, a knot of anxiety amidst the excitement and anticipation I felt to be bound for the mountains of my mind.

Arriving in N'Djamena, we were met off the plane by Baba Mallaye, Stefan's long-time partner at Chad's National Centre for Research and Development.[16] Joining us were botanist Frank Darius, archaeologist Jan Kuper and filmmaker Srdan Keca. Our transport on to the Tibesti was a twin-engine military plane piloted by a Ukrainian who had ended up in Chad because his eyesight no longer qualified him to fly in his native country. That wasn't a circumstance foreseen in my risk assessment, but we were in safe hands. Half an hour into the flight, I looked out the crazed window in the old Antonov's fuselage: the fawn desert plain was featureless but for arcs of darker scrub that traced out meanders of the dry Bahr el Ghazal riverbed. There was no sign of roads, though here and there the tin roof of a dwelling would momentarily flare with reflected sunlight. Until 5,000 years ago, this vast area is said to have been covered by a 'mega-lake' the size of Germany; the area of present-day Lake Chad is less than half a per cent of what was once here.

After a refuelling stop in the oasis settlement of Faya, we were soon above the Tibesti's south-eastern corner.[17] It was an ineffable thrill to lay eyes on the volcano summits for the first time – even from this distance, through a smeary window and desert haze. Just calling their names out to the others above the din of the engines was enough to make me almost delirious with excitement: Emi Koussi, Tarso Yega, Pic Toussidé . . . I couldn't believe my luck.

★

Stefan (who bears more than a passing resemblance to Peter O'Toole in David Lean's 1962 film *Lawrence of Arabia*) was just eighteen when, in 1970, he bought an old Volkswagen microbus and drove from Munich to Kathmandu. The arid landscapes of south-west Asia captivated him. Later, as a student in Berlin, he chaperoned guests of the city's renowned film festival, including Martin Scorsese, Robert De Niro and David Bowie. But while most would be starstruck amongst such celebrities, Stefan's polestar was by now the sense of freedom he experienced in rocky deserts. 'That's where I feel at home,' he told me, 'and in this place, where nobody thinks anybody ever set foot, you find

Palaeolithic hand axes, Neolithic pottery, meteorites, strange minerals, ancient traces. In the vastness, your existence has no meaning and yet every drop of water becomes unbelievably precious.'

Stefan's footprint in the Tibesti maintains a tradition of German exploration of the region. The first European to visit the massif and tell the tale was a Prussian physician, Gustav Nachtigal, who began his odyssey in 1869 in Tripoli, then under Ottoman control. Five years and 6,000 miles later he re-emerged from the interior of Africa in El-Obeid (Sudan) after the adventure of a lifetime. His assignment was to deliver a gift to the Sultan of Bornu, near Lake Chad, from the King of Prussia. But his real motivation was to follow the example of his hero, Heinrich Barth, 'one of the greatest African explorers of all time', who had crossed the Sahara in the 1850s.[18] Upon reaching a fork in the trail north-west of the Tibesti in June 1869, Nachtigal saw his chance for glory and succumbed to the 'irresistible lure of the unknown'– he diverged from the direct route to Bornu.[19] If only he'd done a risk assessment first.

For days, he and his guides struggled in merciless summer heat across stony plains, and rocky ravines. What little water they carried was evaporating through the goat-skin water bags. None could be found to restock, nor was there any fodder for the camels. Worse still, Nachtigal had a suppurating eye infection and could barely see where he was going. As they drank the last drops of water, the 'fate most dreaded by desert travellers' was close. Too debilitated to walk, the explorers abandoned their baggage and heaved themselves onto the camels. An hour after sunrise, the intensifying heat felt like a 'sea of fire'. Nachtigal's airways were parched, his eyes burned, and he experienced severe strangury. For such a meagre adventure, he could not even draw grim satisfaction from the prospect of posthumous distinction.

On the brink of death, Nachtigal began hallucinating. A colossal horned goat was bounding towards him, a man on its back. As it approached, the goat transformed into a camel; the man into one of his guides, Birsa, who was miraculously bringing two full water bags, collected from a nearby water source, in the nick of time.

Surprisingly unfazed – or at least undeterred – by this brush with death by desiccation, it didn't take long for Nachtigal to find himself in another life-threatening situation. Arriving in Bardai, the main settlement of the Tibesti, he was promptly detained in his tent while the Tubu elders argued over whether or not to disembowel him. After a month in captivity, during which he was tormented daily by girls who pelted him with stones if he tried to sneak out for water, he escaped with his guides during the night. Later, he would have to abandon his rock samples in favour of carrying water. His barometers, thermometers and hypsometers were uncalibrated; his maps, made without astronomical observations, were inaccurate. He later wrote: 'If my sojourn there did not yield the scientific results to be desired, this is explained by the unfortunate circumstances in which I made the journey.' But he is overly dismissive of his achievements, which include many insightful observations of the topography of the Tibesti and customs of the Tubu.

My own experience of Bardai was quite the reverse of Nachtigal's. We were welcomed warmly by the local governor and ushered to a campground next to his residence. Here, we assembled our vehicles and met Mehdi, who would be our guide throughout the journey. We filled up on water and fuel, and bought all the dates we could carry. Our first destination would be the most prominent volcano of the Tibesti, Pic Toussidé. Surpassing 3,200 metres above sea level, it is the second-highest point in the Sahara, and a beacon on the trans-Saharan routes between Libya, Niger and Chad. Dick Grove had reached its summit in 1957 and had written notes on its geology in a paper, in which he added hints of the personal experience: '[The lava flows] are bounded by high lateral ridges of large rough blocks. These can be negotiated without much climbing skill, but are very sharp and abrasive, damaging to boots and knees.'[20] From its unblemished conical form and patina of its lavas, Toussidé appears to be the youngest volcano of the Tibesti. In space images, it looks very much like a black octopus – the tentacles formed from snaking lava flows, the head an apron of lava confined within the subdued rim of an ancient caldera.

Our route to Toussidé followed closely that taken by Nachtigal as he fled Bardai. Crossing the volcanic plateau, his feet were cut by sharp rocks, and he shivered at night since most of his garments and bedding had been stolen. Four-wheel-drive vehicles and sleeping bags made for an easier journey by far, but the going was still slow owing to the rough ground. We only made it about fifteen miles beyond the oasis before pitching camp in a ravine of orange-pink ignimbrite that put me very much in mind of the Altiplano in Chile, with its knife-cut *quebradas*. The Tubu describe this landscape aptly with the word *tarso*, which translates as a steep-sided plateau, 'difficult to reach but easy to traverse'.[21]

It was late the following afternoon as we approached Toussidé across an intricately gullied stony plain dotted with ancient rock tombs and stone circles. Each wrinkle in the topography was accentuated by a drape of pink pumice, the dregs of the last large explosive eruption here. The looming volcano was precisely demarcated by its skirt of lava flows, as if an enormous quantity of pitch had spread around its base. We set up camp right at the toe of this endless bulwark of jagged black rock. As the sun set and the air chilled, we explored an Ottoman-era fort with a wide enclosure and sentry posts.[22] Situated well over 2,000 metres above sea level, on the route to Zouar, the Tibesti's second largest outpost, this would have been a strategic spot.

We were up at dawn to climb Toussidé, eager to get going while it was cool. As the sun rose, the volcano glowed an intense brick orange, with sinuous shadows tracing the paths of untold lava flows. Once we had clambered up the flow front, we faced an immense rampart of blocky lava, like an endless accumulation of washing-machine-sized shrapnel. It looked somewhere between daunting and impenetrable. Each step had to be perfectly placed to avoid a twisted ankle or worse. There was no soil, no vegetation. I stopped often to film the scene and soon fell behind the others.

The only way up was via the lava channels, which were separated by high barricades of loose rock. These ducts invariably culminated in an unscalable cliff, with the only option being to clamber up the crumbling side walls in the hope the adjacent channel would lead further. Again and again, I found myself having to cross the grain of the volcano – just as Dick Grove had described it. Reaching another precarious ridge, I paused for breath, and smiled when I saw it was the last. On the other side was a steep tract of pink pyroclastic gravel. But disconcertingly, there was no sign of the others. Had they got that far ahead? Then I spotted a figure near the summit, leaning forwards, as if looking through binoculars or taking a picture. I was sure it was Stefan and waved, but he did not respond. A moment later, he was gone. I pressed on, but the gravel rested on firmer rock and I might as well have been trying to climb a tower of ball bearings, my purchase on the ground all the less controlled thanks to an unwieldy backpack, camera gear and tripod. I could only keep going on all fours.

Seven hours after setting out and having taken little rest, I reached the peak, elated. But Stefan was nowhere to be seen, nor any of our team. I was baffled. Had I been hallucinating, like

Nachtigal? Had they dropped down the other side of the cone to find a more sheltered spot to camp? In any case, the best option was to stay put. I only had half a bottle of water left and didn't need further exertion. To add to my concern, there was no answer when I tried reaching Stefan by satellite phone.

The summit crater was roughly the size of a soccer field, with a sloping rim of toughened pink ash on one side and a mound of dark lava on the other. The floor was sparsely vegetated with sagebrush, sedge and grass. There was not a single trace of human presence – no rusted tins, no hearths, no stone tools or cairns, no graffiti. Several steaming grottos in the lava nurtured terrariums of ferns, herbs and liverworts. I stuck my hands in one to warm them up. Meanwhile, the setting sun cast a shadow of Toussidé's perfect cone across the *tarso*, pinnacles and castellated ridges poking above desert haze. In the near distance, the gaping caldera of another volcano, Trou au Natron, was filled with a tenuous mist. I suddenly felt compelled to reach for the phone again, this time to call Dick Grove at his home in Cambridge. The line was broken, but I think I managed to explain how he'd inspired me to follow in his footsteps to the top of Toussidé, nearly sixty years after his scramble up the mountain.

By twilight, I spotted a torch flash at the base of the summit cone. I shouted, and to my great relief, established contact with the others. I couldn't understand how I had got ahead, but it made no sense to descend in the dark. After a meal of dry-as-dust dates, I slid into my sleeping bag. I needed it – the air temperature fell well below freezing. It was so still, all I could hear was my heartbeat.

In the morning, I scurried down the cone. Stefan was overjoyed; he'd feared I might have broken a leg amidst the lava fields. I explained sighting him at the summit and he laughed. 'Maybe you saw a barbary sheep.'

'But it was taking photographs!' I protested.

Clearly I was dehydrated, and so looked for water to refill my bottles. But there was none; the guides had brought far less than we'd agreed in camp. This was devastating news, as the descent on this ground was unlikely to be a breeze, especially with my extra burden of rock samples.

Had the guides not found a far easier route down, I would have been in real trouble. But the heat only increased as the morning wore on and as we lost altitude. I took a few drops of my remaining water every twenty minutes. Feeling increasingly debilitated, I struggled to keep moving. The infinitesimal relief each next dash of water would bring was my incentive. It was after dusk when we reached the vehicles, seven hours later. Though I wasn't near the state that Nachtigal had found himself in, this remains the closest I've been to reaching a standstill. Looking back I had put myself and my colleagues in danger by abandoning three key principles of exploration: time spent in reconnaissance is rarely wasted; don't get separated from your group; and never, ever take it for granted that someone else is bringing your food or water.

★

While Toussidé has archetypal volcano form – it is classically conical and covered in lava – Trou au Natron, our next destination, is a gaping orifice, and a very large one – five miles across and more than half a mile deep. Peering into the maw, it was hard to believe there was anything but a freefall route to the bottom. But Mehdi knew of a trail cut into the jointed cliffs of beige-pink ignimbrite that formed the upper strata of the volcano. The crater floor, veined with dry watercourses, appeared so far below I had the impression of looking down from an aeroplane window. Patches of white, pink and grey soda deposits shimmered at its centre, punctuated by three dark-brown cones – one of them perfectly symmetrical, like an open parasol.

We came to a small cavern in a grey-pink band of lava; it had clearly been hewn out artificially. The path was strewn with fragments of grey obsidian, and I realised this was a prehistoric quarry, where seams of the vitreous rock had been mined as raw material for stone tools. It was an extraordinary and unexpected vestige of the Tubu's ancestral culture. Further down, we reached a shelf in the caldera wall scattered with patches of chalky white rock. These deposits were double my height and distinctly layered. They sat on the same lava rubble we were crossing, and were evidently vestiges of a much younger veneer. I would have taken them for volcanic ash, but Stefan, already at the outcrop and examining a piece with his hand lens, called out that they were 'diatomites'. So, not an igneous rock at all, but a biological one: diatomites are accumulations of the miniscule silicic fossils of diatoms, a variety of freshwater algae found in lakes. This was astonishing because we were still 500 metres above the crater floor – the implication was that a lake at least that deep and nearly four miles across had once existed here. That is a phenomenal amount of water to contemplate in the middle of the Sahara.

Stefan read my thoughts: 'It is impossible to believe there was such a big lake here, even in the green Sahara time.' Today, annual rainfall in the Tibesti mountains amounts to a few centimetres, a fraction of what Las Vegas, one of the driest cities in the U.S., receives. Even had there been ten times as much rain in the past, it would have taken centuries to fill the crater. And with evaporation equivalent to a centimetre or two of water *per day*, how could any rainfall ever accumulate? On the other hand, it was hard to see how these could be mere localised sediments – there was nothing to hold water here except the cliffs on the opposite side of the caldera. The only alternative explanation is that the crater has deepened substantially since the lake was present, leaving the diatomite high and dry. But that, too, seems far-fetched.

Stefan, Frank and Jan set about taking samples, while I imagined a scene like present-day Crater Lake in Oregon, with obsidian miners coming to the shoreline to fill their water containers. Continuing along the trail, Mehdi located a *guelta*, a freshwater pool, in a deep cleft in the rockface. Shrikes flitted at the water's edge as we filled our bottles using a bucket on the end of a rope. I'd not go thirsty on this volcano.

Over the course of our expedition, I'd heard Stefan frequently refer to the Tibesti as 'the water tower of the Sahara'. After we had set up camp on the crater floor, I got the chance to probe what he meant by that. He explained that 10,000 to 5,000 years ago during the green Sahara period, the Tibesti massif, because of its great height and extent, drew rain from monsoonal air. Over the millennia, this charged a vast aquifer. The largest lake of the Sahara is 250 miles away, where there is no rainfall now, and where the evaporation rate is equivalent to six metres per year. Many thousands of people flock to the oases there for work in the date-harvesting season. 'The only explanation for such unbelievable amounts of water,' he explained, 'is the rain that fell on the Tibesti thousands of years ago, that was stored beneath the mountains, and still flows underground. If the volcanoes never existed, all you would find here would be baking sand where no human set foot for 5,000 years.'[23]

We were back on the road for the next week, heading for the east side of the massif, and hoping to explore Tiéroko volcano, of which next to nothing is known. Wilfred Thesiger described it as 'the most magnificent of all the Tibesti mountains', but even Britain's most indomitable twentieth-century explorer could not reach it. The only visit by outsiders that we knew of had been made by three mountaineers on Dick Grove's expedition in 1957. We were really going off the map, and I probably felt the same kind of wild excitement Nachtigal had experienced as he headed into the unknown at the fork in the trail a century and a half earlier.

Along the way, if I spotted an intriguing rock outcrop from the car window, I would ask to stop, dash out with my hammer, bag a sample, take a photo and GPS reading as quickly as possible, and jump back into my seat. Stefan's approach, meanwhile, was both systematic and opportunistic, since he continuously recorded our position, while frequently taking photographs and writing notes. He often consulted a photocopy of the copious fieldnotes he had written on his previous mission to the Tibesti (like Oscar Wilde's Lady Bracknell, perhaps he enjoyed having something sensational to read on the long journey). It was very impressive practice – Stefan wouldn't even go for a pee behind a sand dune without taking his camera and notebook, just in case he might find a curiosity. Every bird in flight, each rock engraving or unusual landform, was imaged and logged. I am always advising students to work like this in the field; you never know if you will get the chance to return, and I have often kicked myself, months after a volcano trip, when a pattern starts to emerge in the data but there is a missing piece of the puzzle – a rock sample I could have collected, or a lacuna in my gas measurements just when things were about to boil. My students laugh when they see my sample bags with cryptic labels like 'from near the place we had lunch' instead of GPS coordinates.

Each day on our journey, we would set up camp wherever we'd got to by sundown, when it was time for our Chadian colleagues to observe *maghrib*. One evening, we camped close to a particularly large rock tomb at the foot of a bluff of ignimbrite. This was the far northern limit of a vast shield of tuff spread around Voon volcano, whose caldera is five times larger even than that of Trou au Natron. I clambered up the rockface and realised we were at a confluence of two dry valleys. There was abundant evidence that Neolithic people came here: everywhere I looked on the ground, there were pottery fragments and worked flakes and chipped cobbles of obsidian and other rock types. Jan, visibly excited by the archaeological potential, described it as a 'lithic workshop'. 'This would have been a strategic location to watch animals and people following the watercourses,' he explained. 'And this isn't just somewhere people stopped overnight; this is where they produced tools and likely stayed for long periods.' I collected some of the obsidian pieces, wondering if some might prove to match the raw material we had seen in the crater wall of Trou au Natron. Once again, I felt we'd come to a portal between present and deep past. None of these symbols of human presence and mobility would be here but for the volcanoes – without them, this would just be another barren tract of sand.

Despite the long days of driving, the itinerary continually presented new wonders. I sometimes felt I'd passed through several countries in a matter of hours – from wadis filled with tamarisks and acacia with clumps of *cistanche* in yellow bloom at their trunks, to basalt plains and stony desert. Sometimes the scenery seemed straight out of photographs taken by NASA's Mars rovers. The extra-terrestrial reverie might then be abruptly punctured at the sight of a rusting tank, a relic of the Chad-Libya conflict, its gun barrel aimed impotently at featureless haze.

The largest community we saw on the eastern side of the Tibesti was Yebi Bou, still pockmarked with bomb craters from the war. One of our drivers hailed from the village, and we paused so he could visit his family. The cluster of oval dwellings, made from mudbricks and palm mats, overlooked a gorge filled with date palms: another oasis fed by the Tibesti's 10,000-year-old fossil water.

The village had been the starting point for the three alpinists who surveyed Tiéroko in 1957. They had travelled by camel, adept in such a landscape as they are able to step delicately between rocks. It was going to be trickier to get our unwieldy vehicles to the caldera. The mountain is twenty miles across at the base, but lacks the conical form of a pristine volcano, its flanks being scored by deep ravines, or *enneris*, and its rim defined by serrated ridges, sheer-faced towers and toothy pinnacles. The climbers had given them names like Hadrian's Wall and Battleship – but these were inept monikers for the imposing ramparts that soon confronted us.

We had to hike the last stretch up and once inside the caldera I felt I was in the sandpit of giants. I set about covering as much ground as possible and collecting the most interesting rocks. The state of erosion and immense banks of layered gravels reminded me of a formerly glaciated landscape. Could ice have once capped the volcano, I wondered? I wish there'd been more opportunity to explore, but we were now under time pressure, with more than a week's driving, all the way back to N'Djamena, ahead of us. We had achieved what we'd set out to do, but I felt almost bereft at having to turn around so quickly.

Our route bore south, through the Miski valley, the great whaleback profile of Emi Koussi in the distance. After a near miss with an abandoned minefield (the only time I saw Mehdi lose his cool), we reached the Sahara of my imagination – dunes, Fata Morgana, torrid by day and frigid at night. Some distance beyond the settlement of Faya, we spied an empty palm-thatched shelter and took the chance of a rest in shade. Within minutes, overcome by the torpefying heat, all my companions were asleep. One of the drivers lay motionless within the splinter of shadow cast by his Landcruiser. Anyone stumbling across the scene might have suspected a catastrophe.

Not much later we encountered one in the making. We had not been back on the route long when we saw something inconceivable: two faltering figures crossing the sand on foot. We drove up and found them to be young men carrying a couple of empty detergent bottles. They were gravely dehydrated and explained they had broken down and were looking for a water hole. We gave them water and went in search of their car, which was several miles away. It astonished us that they had managed to walk so far in the middle of the day. But more staggering, they had come to this desolate place in a Kia compact city car. It was tipped forwards, its front end embedded in sand. To the side was a boy of ten, suffering from heat stroke, and an elderly man. We gave them water, and asked why they were out here, risking all. They explained they were fleeing Libya, hoping to reach N'Djamena and then Europe. They had several days' driving still ahead. We pushed the car backwards and slid traction boards under the tyres to get past the dip where it had grounded. They were freed but ten minutes later, had snagged in the sand again. We extracted them again, and so this continued until we reached the safety of a definite trail the next day.

It is very likely the refugees would have perished had we not come along when we did – we saw no other vehicles and there is no clear route across the sand. I cannot imagine how desperate their circumstances in Libya must have been for them to contemplate such an ill-equipped journey. I recalled Stefan explaining to me the choices faced by Neolithic people as the Sahara expanded again 7,000 years ago. 'Survival means mobility,' he had said. The vignette we had witnessed of the flight and plight of contemporary migrants seemed like an echo of ancient passages via the crossroads of the Tibesti.

★

Thankfully, unlike Dick Grove and Gustav Nachtigal, I didn't mislay or abandon my rock specimens. The work on them continues, and we have already made fresh discoveries and are understanding more about the origins of the Tibesti. Several of the pyroclastic samples have been dated, giving a sense of the history and longevity of volcanism here – Voon's caldera formed eight million years ago, while the pumice drifts near Pic Toussidé are the traces of an eruption that took place just 100,000 years ago. The meaning of Toussidé – 'which killed Tubu people by fire' – hints at much more recent eruptions that occurred within the span of oral history.[24]

Another puzzle I wanted to solve was how this volcanic region had formed so far from any tectonic plate boundary. From analysis of the lavas, it appears the answer lies in an uncommonly hot region of the mantle under the Tibesti, and localised stretching of the plate, which provoked rock melting by releasing pressure.[25] The resulting magma gathered in chambers in the crust, sporadically feeding eruptions large and small.

Meanwhile, geochemical fingerprints of the obsidian I collected connect the artefacts from near Toussidé, Bardai and Tiéroko, revealing that the timeworn trails we followed across the massif have been in use for many thousands of years. I suspect Tibesti obsidian was prized even further afield – in Sudan, southern Libya, Niger, and perhaps as far away as Egypt – but it will need new field surveys and lab analysis of any finds to prove it.

Stefan's dating of the lake deposits from Trou au Natron shows that water filled the crater between 8000 and 3500 BCE, firmly within the 'green Sahara' period.[26] But we remain baffled by the freshwater sediments found so high above the crater floor – scientific questions often elude easy answers.[27] I continue, though, to marvel at the idea of a vast lake in the middle of what is now the largest desert on Earth.

Following our expedition in 2016, violence escalated significantly in the Tibesti, with deadly confrontations between armed non-state fighters and government forces, and between rival gangs of artisanal gold miners. In 2021, an insurgent group, FACT, based in the Tibesti region, began advancing south. The Chadian president, Idriss Déby, was among those killed on the frontline in the ensuing combat. During our climb of Toussidé, I had remarked to Mehdi that it was sixty years since Dick Grove was there. He had replied that it might be another sixty before the next researcher would come. I had taken it as a joke, but with hindsight, it may well be a realistic prediction. This weird and wonderful land is likely to remain very seldom seen by the outside world for the foreseeable future. I thank my lucky stars I got to see it when I did.

I had gone to the Sahara seeking rocks and stratigraphic sections, but our expedition opened my gaze not only to a spectacular, extra-terrestrial landscape, but also to the vital significance of the mountains for human life. Had the Tibesti never existed, there would be no rain, no aquifer, one less haven in the desert, and the eastern Sahara would be much harder to cross. The *tarsos* at first appear desolate, devoid of human traces, yet look closely and it becomes clear that mounds of boulders are actually prehistoric tombs; scratches in cliff faces are engravings of cheetahs, elephants, cattle; and flecks of glassy black rock glinting in the sand are Stone Age tools. This human footprint, along with the evidence in the rocks, such as the lake sediments, for cycles of expansion and contraction of the Sahara, affirms the transience and persistence, the fragility and mobility of life. And it all plays out where extraordinary volcanic forces prefigured environment and habitability, above all by attracting and storing abundant fresh water. The desert appears empty only to those who do not understand it.

Flame in a Sea of Gold

At the edge of the crater, Antarctica.[1]

'... scientists walk the boundary between sensory, intimate experience and objective, universal knowledge as embodied people in the world.'

— Jessica O'Reilly, 'Sensing the ice'[2]

Mount Erebus is unpopulated. No flowers bloom on the Antarctic volcano, no mice hide in the rocks; its icy form baffles the flow of Earth's purest air while adding to it a stream of its own aged brew of effluent. Ice sheets waxed and waned during a million years as its stature took on the world's largest volcanoes. No rain falls, no groundwater flows; only the hiss of its primal lava lake and strange twang of icequakes break the silence of a still day . . .

It is a lonely place but for me it brings an intense joy: an intoxicating solitude that comes from immersion in an unfathomable icescape.

The geologist Tannatt Edgeworth David, who was the leader of the first party to peer into the volcano's sibilant maw in 1908, exalted its 'incomparable grandeur and interest'. For him, Erebus was among 'the fairest and most majestic sights that Earth can show'. Having spent a year of my life living near the volcano's summit, plus collective years on a hundred other volcanoes, I can say Erebus *is* the most sublime and imposing sight I ever beheld. More people have been in space, and many more have summited Mount Everest, than have looked upon Erebus's fire pit.

Though Mount Erebus is new to myth-making, the mythical associations of its name tie it to human imagination of the distant past and the land of gods. In Greek legend, mighty Erebus was born of Chaos, and associated with the beginning of time, with darkness and the shadowy realm of the underworld.[3] Thanks to Romanticism, this symbolism grew: Erebus evoked literary or metaphoric blackness, the abyss, the afterlife. Following the volcano's discovery, the idea of a lofty and sublime Erebus, of fire and ice, ignited the public imagination of Earth's extremities. All this proves the inspiration in so naming the mountain, which came about through a quirk of the British Navy.

The volcano was discovered in 1841 during the Antarctic voyages of Captain James Clark Ross, whose mission was dubbed 'by far the greatest scientific undertaking . . . the world has ever seen' by the polymath William Whewell (who coined the term 'scientist').[4] That same year, Franz Junghuhn was exploring the territory around the Toba super-volcano in Indonesia, Sartorius von Waltershausen was mapping Mount Etna in Italy, and Charles Darwin was working on his books about coral reefs and volcanic islands. It was also the year that volcanology began as a modern science, with the founding of the observatory on Vesuvius, an exemplar of the era's systema-tisation of scientific training and the gathering of data through institutions. The affinity between the Antarctic and Neapolitan volcanoes goes deeper still, since both erupt a particular variety of magma known as phonolite or clinkstone, owing to its bell-like ring when struck. They are more like distant cousins than twins, though – Vesuvius was born yesterday in comparison and has a lot of growing to do to reach the prominence of Erebus.

Ross's Antarctic expedition came about after the Napoleonic wars, when Britain retained the world's most powerful naval force but lacked an enemy to engage. That freed up its singular resources for an era of exploration.[5] The mission's motive was to improve navigation at sea by compass, since the vagaries of weather meant it was often not possible to chart a course by sightings of sun or stars. It was thought that new observations of the Earth's magnetic field in the Southern Hemisphere would establish a theoretical framework for terrestrial magnetism along the lines of Newton's law of gravity. A veteran of Arctic exploration, Ross had been first to reach the north magnetic pole and was desperate to bag its southerly counterpart. He was given overall command of the expedition, two warships – *Erebus* and *Terror* – and a hundred pages of instructions from the Royal Society of London on how to collect the data required.

But why *Erebus*? Why *Terror*? A ship bound for frozen seas had to be sturdy enough to overcome pack ice. Such strength was found in the navy's 'bomb vessels', so-called because they were built to withstand the recoil from firing mortars. Refitted for scientific endeavour and capacious enough to carry boats and provisions, these ungainly but unshakable craft stood the best chance of reaching the ends of the world. Reflecting their original purpose for shore bombardment, the Royal Navy routinely gave them explosive or hellish names, like *Strombolo*, *Aetna*, *Volcano*, *Blast*, *Convulsion*, *Infernal* and *Belzebub*. The abyssal connotations of HMS *Erebus* implied where those in her gunsights were headed.[6]

Each ship had a full complement of sixty-four men as they left England at the end of September 1839. Over the next year, they called at a succession of oceanic volcanoes during a winding voyage south: Madeira, Tenerife, Cape Verde, Trinidade, St Helena, the Kerguelen Islands. If only Ross had had a competent geologist with him – someone like the pioneering volcanologist George Poulett Scrope, perhaps.[7] This would have been such an extraordinary opportunity to study a variety of volcanoes. Instead, the role was taken by Robert McCormick, a mediocre naturalist and surgeon who had sailed with Charles Darwin on the second voyage of HMS *Beagle*, and a man far more interested in ornithology than rocks. McCormick's biographers have found little to celebrate his intellect or character, but one has to acknowledge the genius of filling his geological collection with stones recovered from the crops of birds on his dissecting table.[8]

During a long stopover in Hobart, Tasmania, Ross learned that French and American expeditions had sailed south the previous summer on *the same* meridian he'd announced *he* would take. But this ignominy would turn out to be a stroke of luck. At this time, almost nothing was known of Antarctica save for some scraps of coastline spied by James Cook and a few sealers. So heading into the southern ocean and hoping to find something was like playing the game Battleships: the polar map was an act of imagination. Since it would be 'inconsistent with the pre-eminence [England] has ever maintained' to follow others, Ross decided to go south on a much more easterly meridian (170°E, which cuts through the South Island of New Zealand). The gamble paid off, since, on this trajectory, Ross was able to sail much further, owing to the vast embayment in the Antarctic continent later named after him: the Ross Sea.

The ships encountered pack ice soon after New Year's Day 1841, sustaining 'very heavy blows' that could easily have sent ships and men to the seabed. They became trapped until the wind slackened and the ice loosened. Then, quite suddenly, they broke through and entered a vast sea with not a particle of ice ahead. Ross, his excitement at fever pitch as the lines of magnetic force steepened, steered towards the estimated position of the pole. But two days later 'land-blink' dashed his hopes. While most would have rejoiced at the sight of the yellowy gloom that signals looming land, for Ross it forestalled a direct passage to the pole under sail. As they drew closer, a mighty range of glaciated mountains came into view, and at least he had the satisfaction of naming its peaks and running up the British flag to take possession of some islands on behalf of Queen Victoria (bemusing the penguins in the process).

Southerly winds and tides slowed their advance, and sea spray froze to the bows and running ropes, but at noon on 27 January, further south than anyone had ever been, the crew sighted a

'High Island'. At first, Ross mistook the vapour trail emanating from the summit of the island for snow drift. But while the men took sightings, a 'dark cloud of smoke, tinged with flame' burst from the summit, rising in 'a perfectly unbroken column, one side jet-black, the other giving back the colours of the sun'. By now, all officers and crew on deck were transfixed by the sight of a towering, ice-clad, active volcano. For Joseph Hooker, the young botanist aboard the *Erebus*, it was an awesome sight 'surpassing everything that can be imagined'.

Until now, Ross had named topographic features after luminaries of the Royal Society, the admiralty and government, as well as benefactors and friends. But this volcano he named after his ship, and its extinct neighbour after the *Terror*. Perhaps naming a smoking volcano after a Lord of the Admiralty or the Queen could have been misconstrued and cost him a peerage. Maybe he was inspired by the ships' volcanic associations – *Erebus* was a *Hecla*-class vessel (Hekla is the Icelandic volcano that Robert Bunsen investigated), while *Terror* belonged to the *Vesuvius* class. Or was it the mythological and plutonic associations of Erebus and Terror (Diemos) with the underworld, darkness and threat? Whatever the commander's motivations, the ships had surely earned the honour after their colossal circumnavigation.[9]

Ross still hoped to find a way to sail closer to where the magnetic pole should be, but the coastline was unbroken, nor could he get through ice pack to reach a natural harbour on the mainland where he might overwinter.[10] There was no option but to retreat north.

Even Ross knew few would understand his 'deep feelings of regret' at being thwarted in reaching an aspect of the globe beyond human sensibility, one signified only by a vertically dipping magnetised needle. It's probably a good thing he didn't know the magnetic poles are not even stationary. In any case, what he discovered is far more significant than he could ever have dreamed. The Transantarctic Mountains and West Antarctic Rift System (which rivals the East African Rift in scale) are features of global significance for understanding magmatic, volcanic and tectonic processes. The island on which Erebus rises is now named after him, and home of the largest scientific hub in Antarctica. Captain Robert Falcon Scott later said of Ross's achievements: '[he] wrested an open sea, a vast mountain range, a smoking volcano, and a hundred problems of great interest to the geographer'.[11] This success was born of experience, good planning, an open mind, and a good dose of audacity, valour and resolve. Sure, Ross had a huge team behind him on board the vessels and back in Britain, and he got lucky – but he also made his luck.[12]

*

Many of the great explorers, like Ross, were military men. I wasn't cut from such cloth. Perhaps closer to my mark is the Welsh-Australian geologist Tannatt Edgeworth David, who became known by his comrades as 'the Professor'.[13] He was a member of Ernest Shackleton's *Nimrod* expedition when he trekked to the top of Erebus in 1908.[14] What impresses me about David is his proclivity, even in late career, to change tack at the drop of a hat. Imagine: Shackleton learns of the Professor's reputation, tracks him down in Sydney and says something like, 'Do you fancy joining an Antarctic expedition for a month or so? It will take a while to get there, and I can't guarantee plain sailing, but you'll be back by next semester.' David accepts the invitation. Then, on the voyage south, during which David turns fifty (on the day Erebus is sighted), Shackleton ('the Boss') persuades him to join the shore party rather than return with the ship. David agrees, and writes a letter to his superiors requesting exceptional academic leave, knowing his decision is a *fait accompli* since he cannot possibly receive a reply, favourable or otherwise, for another year.[15]

What I also admire in David is his resilience and honesty. Though the oldest man of the shore party and a respected university professor, he did not shirk manual labour or extreme footslogs, but he also didn't try to hide the rigours with a smile for the camera. There is a famous photograph of David and his companions at the south magnetic pole. It was a terrible journey both there and back, and almost utterly pointless. You can see David holding the thread to activate the camera's shutter. He looks comprehensively shagged, half human, and somehow cognisant that he had endured it all only to reach a point of no permanence or meaning.

The *Nimrod* reached Ross Island in February 1908, and the shore party threw themselves into building their winter quarters at Cape Royds on the western fringe of Erebus. It was a choice spot near penguin and seal colonies, which meant a ready supply of food and fuel. And it afforded a spectacular view of the ice-gripped smoking volcano, whose allure was potent. David wrote: 'For us, living under its shadow, the longing to climb it, and penetrate the mysteries beyond the veil soon became irresistibly strong.' On 4 March, just a month after landing, Shackleton yielded to that urge, and the next day a party of six set off: David and his protégé Douglas Mawson (a geologist, but serving as the expedition's physicist), Dr Alistair 'the Mate' Mackay (assistant surgeon), Lieutenant Jameson Adams (meteorologist), the multi-talented Dr Eric Marshall (surgeon, surveyor, cartographer and photographer) and the twenty-year-old Sir Philip Brocklehurst (appointed assistant geologist by Shackleton, though untrained in the subject and valued more by the leader for his physical strength and his mother's wealth). Another young geologist with the shore party, Raymond Priestley, was left behind. He probably felt peeved, but his chance to summit would come.

The party harnessed up to the sledge, which was piled high with instruments and supplies, and set off from Cape Royds at 8.45am. They skirted crevassed ice by following rock outcrops and moraines, but soon encountered a vexing tract of glassy blue ice thinly disguised beneath soft snow. Then, as more snow began falling, they hit a stretch of *sastrugi*, a stiff-peaked meringue of wind-sculpted snow, all the more difficult to cross with a ponderous sledge in a whiteout. The men cursed freely as they slipped and tripped; feet scrunched the ice while the sledge runners sawed it. But when they paused to rest, and once they had got their breath back, the silence was intense.

They made their first camp near an outcrop of lava. Starving, they tucked into their first bowls of hot, greasy 'hoosh', an Antarctic staple consisting of boiled pemmican (dehydrated beef and fat), chipped Plasmon biscuit (made with dried milk) and emergency rations. The Professor, more accustomed to polite meals in the faculty hall at the university, recorded that: 'A man after such a meal, in any but a polar climate, would have seen in his sleep "more devils than vast hell can hold".[16]

One man's devils would likely have been another's – it was three to a sleeping bag. Whenever one stirred, all were awoken by showers of tiny ice crystals formed when their breath froze on the reindeer-hair lining the bags. It was -23°C (-9°F) outside the tent. After more hoosh for breakfast, it was time for another trudge across *sastrugi* – but, with the gradient steepening, the effort was even more back-breaking. The sledge capsized repeatedly and had to be repacked – hot work, despite the freezing air. By late evening, they pitched camp amongst 'fresh volcanic slag', indicating recent explosive eruptions. The thermometer read -33°C (-27°F) overnight.

ERUPTIONS

Clockwise from top left: Vesuvius (Italy) with lava flows in 1756, from William Hamilton's *Campi Phlegræi*. Stromboli (Italy) snapped from a safe distance in 2008. Lava dome atop Merapi (Java, Indonesia), from Franz Junghuhn's *Java Album*. *Nuées ardentes* descending Soufrière Hills (Montserrat) and feeding ash clouds.

CREATION

Top left: Champagne Pool at Waiotapu (New Zealand) is acidic and 75°C,
yet teems with microbial life, as do the fumarolic ice caves of Erebus
(Antarctica), below. Top right: Fertile volcanic soils near Mount Sinabung
(Sumatra, Indonesia).

DESTRUCTION

Above: Advancing lava during 2021 eruption on the island of La Palma (Canary Islands). Below: Photographs taken a year apart from the same spot on Soufrière Hills (Montserrat) – note slope on right for orientation. The church pictured was obliterated by *nuées ardentes*.

GLOBAL REPERCUSSIONS

Clockwise from top left: Sulphur (here being mined on Ijen volcano, Java) is the volcanic effluent that can change global climate. Sulphurous dust from the 1883 Krakatau eruption caused dramatic sunsets, as seen in this drawing made beside the River Thames in London. The 'blue ring' in this stained section of a pine tree signals very cold late summer weather in Central Asia in 536 CE (photo courtesy of Ulf Bŭntgen). The author (at his sweatiest) inspecting 74,000-year-old ash deposits in India sourced from the Toba super-eruption.

IN THE FIELD

Clockwise from top left: The author, singed on Momotombo (Nicaragua) and frozen on Erebus. The author with an infrared spectrometer on Masaya (Nicaragua). A colleague of the author's installing a bespoke radar to measure lava lake height at the edge of Halemaʻumaʻu crater (Hawaiʻi).

ORIGIN STORIES

Top left: Stairway to 'Heaven Lake' on Paektu volcano, Korea's ancestral mountain. Top right: Ancient rock engravings on lava at Pu'uloa ('hill of long life') in Hawai'i connect genealogy with geology. Below: Ruapehu Crater Lake (New Zealand) brought life to Mother Earth, according to local Māori belief.

TRACES AND ARTEFACTS

Above: Trou au Natron volcano (Chad) hosted an ancient lake and obsidian quarry. Bottom left: Centuries-old footprints in ash on Kīlauea volcano (Hawaiʻi). Bottom right: Obsidian implements the author found on Nabro (Eritrea).

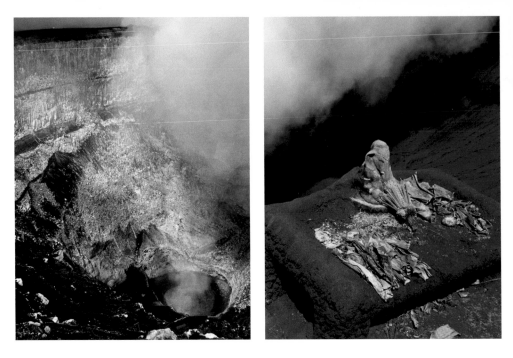

PORTALS TO OTHER REALMS

Top left: The animated lava lake of Ambrym volcano (Vanuatu). Top right:
Offerings beside Bromo crater (Indonesia). Below: Easter Islanders considered
Ranu Kao crater the entrance to the underworld for the souls of the departed.

The next morning brought a change of plan – only David, Mawson and the Mate were supposed to summit, but having made it this far, it was decided all would go for it. But there weren't enough crampons to go round, so they had to cut steps into the ice. Feeling the thinning air, they stopped for the night near a rocky ravine. So far, the weather had been kind, but as they slithered awkwardly into their sleeping bags, and pulled the tent canvas over the top, menacing cloud was gathering. In an hour, the wind was howling, filling their sleeping bags with snow.

The blizzard still raged in the morning and the party had to stay put. Brocklehurst and the Mate got up in the afternoon, perhaps for a pee, and were instantly knocked down by the force of the gale. Marshall, now their sleeping bag's only anchor, struggled not to vanish like his companions until they rematerialised from the whiteout on hands and knees. Brocklehurst was in a bad way, but a little biscuit and Bovril chocolate lifted his spirits. There was nothing to wash it down, as it was impossible in the lashing wind to light the stoves to melt ice.

At around 4am, the atmosphere calmed. They got the stove burning, ate a hearty breakfast and pushed on after sunrise. Their skyline was the rim of an ancient caldera, behind which the summit cone lay. At last, they reached the crest and gazed upon the steam-crowned summit crater of Erebus they had so often marvelled at through the telescope from below. Also prominent were dozens of curious ice mounds. Later, they realised these formed where steam emissions from the ground froze in the air, forming bizarre ice sculptures – one looked just like a lion. Frozen fumaroles like this had not been documented before. Recent studies on Erebus have revealed they are biological refuges where bacteria and fungi flourish in the volcanic heat and gases. Some ecologists think they might provide clues as to where else in the solar system life might lurk.[17]

The exhausted party clambered down the caldera wall, pitched camp in a rocky gully and prepared lunch. Brocklehurst confided that he'd lost feeling in his toes. Marshall inspected them, and found several had turned black with severe frostbite.

The morning greeted them with the spectacle of the bluish-grey shadow of Erebus cast on a 'rolling sea' of golden cloud far below – a scene of such 'transcendent majesty and beauty' that David was lost for words. Although the route to the summit looked easy, the altitude made for a 'painfully slow' ascent, except for poor Brocklehurst, who stayed in the camp. At last, the men stood at the crater rim, a 'vast abyss' filled with steam. David observed: 'After a continuous loud hissing sound, lasting for some minutes, there would come from below a big dull boom, and immediately afterwards a great globular mass of steam would rush upwards . . .'

This cycle of frothing and booming continued throughout their time at the crater. They set about collecting rock samples, and taking readings and photographs, but couldn't stay long – they had to get Brocklehurst to Cape Royds. Back at the camp, they ate, packed and started down, following again the spurs of rubbly lava. But the last rib petered out abruptly above a peril-ous ice slope. Too tired to go back up and find another way, they took a gamble with gravity, launching their packs, then their bodies, and hoping to come to rest gently rather than smash into rocks. Somehow, their 'wild career' paid off, though their cook-ware took a beating and some instruments were lost or broken. They reached Cape Royds the following day, where David wrote that the adventure would 'never fade from memory'. Even if it did for Brocklehurst, he would have been reminded by the regular sight of the big toe of his right foot in a jar on the mantel-piece at home.[18]

The haul of measurements, photographs, visual observations and samples that David and his companions brought back from this first ascent of Erebus is extraordinary – it includes descriptions of the ice towers and explanation of their formation; recognition of the similarities between the rocks of Erebus and those of the volcanoes in eastern Africa (an astute conclusion since Erebus, too, is located in a tectonic rift related to a mantle plume); and evidence for explosive eruptions.[19] They had penetrated some of the mysteries 'beyond the veil' of the mountain, setting the stage for all who would follow. Referring to the climb, the historian Edward Larson wrote that it showed 'how science gave meaning to adventure'.[20]

★

On day one of my PhD studies, I had discovered a stack of half-inch computer tapes on my desk. A note beside them read: 'These are satellite images of Mount Erebus in Antarctica. It has a lava lake and should show up in the infrared channels. Take a look – see if you can do anything with them.' Our image-processing system was a large workstation, and reading a tape took hours, so I would mount them in the afternoon, hoping to find the scene loaded to the computer in the morning, rather than a plastic tagliatelle in a heap on the floor.

I had little feel for where Erebus was, and individual pixels corresponded to areas equivalent to several city blocks. Most of the surface was covered in snow and ice, so there was little detail to discern. But the strangeness of the images somehow made them all the more evocative – like glimpsing something extraordinary through a fogged-up window. They transported me to an alien world of my imagination.

What I *could* make out was the irregular shape of Ross Island, and the slopes of Mounts Bird, Terror and Erebus, the latter crowned by a hot pixel or two. These registered the intense thermal radiation emitted from the lava lake. I analysed these data for a few weeks, then had to focus on preparations for my first fieldwork on Stromboli. While my research then took me to fuming craters and cones in Nicaragua and Chile, I couldn't shake an embryonic longing to see the polar volcano in the flesh.

Then, in 1993, the prospect loomed through a chance encounter with a geochemist, Phil Kyle, amongst the frozen volcanoes of Kamchatka in the Russian Far East, where I was part of a joint NASA and Russian Academy of Sciences field campaign. Phil, a professor at New Mexico Tech in the U.S., was there to measure sulphur emissions from the volcanoes with a Cospec, the same device I would use a few years later on Montserrat. He had been going to Antarctica annually since his first geological mission there as a student (in his home city of Wellington, New Zealand), and had lately been using the Cospec to monitor Erebus. As we reconnoitred the area around the beautiful ice-clad Kliuchevskoi volcano, Phil remarked casually that I should join him on the polar volcano someday.

A decade later, he emailed me out of the blue offering me a place on his team with the U.S. Antarctic Program, due to head south in November that year. He had come across my then recent work with a new line of pocket-sized ultraviolet spectrometers. I had shown they could emulate Cospec's function for a fraction of the cost, and Phil was keen to see the devices in action for himself. It felt like my long-held dream was coming true.[21]

Despite my excitement, I had a couple of anxieties regarding the mission. Both came up in discussions on logistics with Phil when we had a chance to meet in Cambridge that spring. We'd already talked for a couple of hours when he looked me in the eye and said, 'You're not vegetarian or anything like that, are you?' Two potential responses flashed in my mind: I could say, 'Certainly not – I think they're weirdos, too,' or I could come clean. I figured it was better to confess than to get there and be served hoosh three times a day. I was relieved when Phil said, 'That's OK – so's my student, and she's keen to take charge of mealtimes.' Then the more alarming query: 'You've had your wisdom teeth out, haven't you?'

I knew the national Antarctic programmes were pretty hot on health and dental qualification for participants owing to the limited medical facilities on the bases, and the complications often encountered when patients needed to be urgently repatriated. My wisdom teeth had not been extracted; they had not even erupted. Truth be told, I hadn't been to the dentist since my teens, but in spite of my phobia I booked an appointment. My apprehension festered, and on the day I almost couldn't walk to the clinic. I know – Brocklehurst and his colleagues would have had a good laugh. But I made it to the padded chair: blinding bulbs, my fists tight balls. To my great relief, the dentist reported my teeth were in good shape, adding, 'And you don't *have* wisdom teeth.' So now nothing could stop me – I was really going to Erebus.

What I didn't know at the time was that after my first trip, I would go back to Erebus twelve more times. That makes me sound hardier than I am – circumstances are very different to Ross's or Shackleton's day and I have never really had to rough it. There were no crossings of the furious Southern Ocean; no man-hauling of sledges loaded with life-support systems and surveying tools up a gigantic volcano; no amputated digits or enforced consumption of hoosh.[22] Instead of weeks at sea, I reached Ross Island in eight hours by military transport plane from Christchurch, New Zealand. We landed on sea ice close to McMurdo station (named after Lieutenant Archibald McMurdo, who had sailed with the *Terror*).

Mac Town as it is locally known, feels like a cross between a bustling frontier town, a university and a cargo-handling facility. In the austral summer it is home to around 1,000 people and an occasional Emperor penguin. Attractions included 'cookie Wednesdays', science lectures on Thursdays and Latin dancing on Fridays. Or to get away from it all, there was the 'Ob Tube' – a vertical pipe with internal rungs providing access to an observation chamber beneath the sea ice where you could listen to the harmonic trills of Weddell seals, once taken for the music of mermaids. There were many amenities, too: a hair salon, fire service, ecumenical church and an ATM if you ran out of cash to spend in the souvenir store or any of the three bars on station. There were dormitory buildings (one known ominously as Hotel California), a huge open-all-hours canteen, which served midnight rations, and a sprawling complex of laboratories and offices for scientists, who were more or less affectionately known as 'beakers'. Helpfully, there were ice machines on every corridor.[23]

At the start of each field season, our Erebus research team would typically spend a week in Mac Town, gathering tents, sleep kits, crampons, radios and provisions. When it came to foodstuffs, if it was dried, frozen or tinned, it was probably available – muesli, bagels, chocolate (two bars per person per day), iced guacamole, pasta, chopped tomatoes . . .[24] A surprising array of options were available once we were out there, and I was clearly not going to suffer as a vegetarian.[25]

Alongside these logistical preparations, we also had to take a series of classes for all of the 'what-if' scenarios we might encounter. There was instruction on survival, in case we became stranded in the field; on recognising and treating hypothermia; on snow-mobile maintenance and safety on sea ice; and on how to avoid decapitation by helicopter rotor. These sessions seemed quite abstract in a warm seminar room, but then, walking McMurdo's dusty streets, often gurgling with streams of muddy meltwater, I would spy Captain Scott's memorial cross atop Ob Hill, or

Discovery Hut, where he had overwintered with Shackleton, and remember where we were.

★

Mac Town is at sea level, where the summer temperature hovers around freezing. It's crowded, and everything that happens is purposeful and punctual or recreational. It is far from exotic. But jumping from the helicopter at nearly 3,000 metres elevation, surrounded by blinding whiteness on the side of Erebus, and taking that first gulp of thin, bracing air at -30°C (-22°F) is when the strangeness envelops and invades. What took David and his men five days, we covered in twenty-five minutes. But that doesn't diminish the joy of being there; it just guarantees you keep your fingers and toes.

We didn't go straight to the summit. Instead, the plan was to spend a couple of days acclimatising to the altitude before heading further up the mountain. At first, the idea of this enforced stay challenged my patience – I just wanted to get stuck in with work. But as the helicopter lifted away to return to McMurdo, and the silence thickened, I took notice of my new surroundings. It was as if we had landed on an ethereal, sculptural, blue-white chiaroscuro exoplanet. The feeling was like no other: I was utterly spellbound.

The temporary camp was located at the head of the Fang glacier, so-called because of the toothy remnant of an ancient crater wall that buttresses it to the north. Three yellow pyramid tents (the one nod to Captain Scott's era), a bamboo stick with a flag marking the communal urinal and a snow-wall enclosure sheltering a bucket awaited us (the field support staff at McMurdo had set up the refuge ahead of our arrival).[26] Snowmobiles were parked nearby with jerrycans of fuel leaning against them. The summit cone of Erebus was hidden from view by the steep shield looming to the south, but its plume of sulphurous gas – my scientific quarry – snaked mischievously overhead.

The rocky Fang ridge overshadowing the glacier bevelled off to a corniced snow ridge, which one could walk right up to. The swoop of ice on the other side gave me vertigo – one slip would start a wild glissade ending in impact with a hard rock nunatak, or a hair-raising chute all the way to the frozen sea. Somewhere down there, too, lay a real impact site: Crash Nunatak. This was hallowed ground, where an Air New Zealand tourist flight had cruised into the side of Erebus in 1979, with the loss of all on board.

In the morning, the air was perfectly still. I savoured the hint of warmth in the sun, which lit up the sea with a brilliant golden glint. The majestic sweep of Mount Terror also gleamed and glistened where facets of icefalls and crevasses reflected the sunlight, making part of the mountain look like glass. It was almost too much to take in. Each time I returned to Erebus since that first season, I found myself looking forward to this period of adjustment at Fang, even if it invariably entailed an unrelenting altitude headache.[27] The slowness of it – offline, and seriously off-grid – signalled the start of reliving the most exciting time of my life.

Lower Erebus Hut, referred to as LEH or simply 'the Hut', our home for the next weeks, was a short but steep snowmobile ride above us. At 3,500 metres above sea level, more than twice the altitude of Denver, it lies within the ancient caldera wall, at the foot of the summit cone, not far from where the Professor and his comrades must have spent the night. There are, in fact, two huts – both squat wood-framed buildings, weather-blasted and bristling with radio antennae. The Hut served as kitchen and living area; the other, 'the Garage', was a workshop and storeroom. The entrances to both were like oversized refrigerator doors – not easy to open against the wind. I'd picked up a NO JUNK MAIL letterbox sign in a store in New Zealand, and stuck it on the front of the Hut.

I was eager to get my first sighting of the crater, but for the first two weeks we were grounded by an unremitting blizzard. But eventually, the wind and humidity dropped; we set out for the summit under a sky so crystalline it felt as if we were touching outer space.

We used snowmobiles to get closer to the crater, but the last stretch was too steep for them so we had to continue on foot – a free rhythm of limbs and lungs, heart beating fast in thin air. A loose bed of grey feldspar crystals, as long as a finger, shifted and scrunched to each footfall. Icicles formed from my humid breath and stiffened my eyebrows, while snotsicles sprouted from my nostrils.

Rounding the crest of the crater at last, I gingerly stepped to the precipice to look Erebus in the eye for the first time. Or better to say 'mouth' as the chasm before me, shaped like an upturned boat shell, was also huffing and puffing. The plunging walls were striped and mottled with snow, pale yellow sulphur and dark bands of rubble freed of ice by steam vents. The horizontal layering intersected with vertical trails of dust that had spilled from scars in the rock face. Sunken into a slope of hardened scree at the far wall of the crater was another shaft, and at the bottom of this I could make out the famed lake of lava. Its surface was rent with fiery scars and rifts, and folded and wrinkled like elephant hide. Bluish fumes streamed out from the seething mass, which looked so stiff and viscous that I couldn't help thinking that anyone who fell in would go *splat* rather than *splash*, meeting a cartoonish end.

When you look at a lava lake you are seeing the top of a column of magma connected to much larger pods and pipes of molten rock deep down. The heat and gases emanating from the lake offer direct clues to that underworld. I knew at once I had found the ideal natural laboratory to study volcanic behaviour. For Sir William Hamilton it was Vesuvius; for Franz Junghuhn, Merapi; for Frank Perret, Montagne Pelée ... I, too, now met my muse.

★

My scientific purpose on this first mission was to measure the sulphur dioxide fizzling from the lava lake. This would tell us how much magma had to be rising up to supply the gas.[28] The answer was about two bathtubs of molten rock per second (these are not our conventional units, mind). The data also helped to understand how Erebus's emissions might impact the pristine Antarctic atmosphere. Back in the 1980s following the discovery of the Antarctic ozone hole, some challenged the view that the ozone loss resulted from manufactured compounds widely used as refrigerants and spray-can propellants. They argued instead that Erebus was responsible. While we found localised ozone depletion in the plume snaking downwind from the crater, there is no way this can account for the continental scale of the 'hole'.[29]

When I came back to Erebus the following year I also wanted to measure other gases exhaled from the lava lake, as these could give clues to the temperature of the magma and depth of its source.[30] To do this, I brought the same infrared spectrometer I had used on Montserrat and with my Icelandic colleagues at Fagradalsfjall. The instrument needs a heat source behind the gases to work, and I was sure the incandescent lava lake would cover that – I would just need to train the attached telescope at its surface from the crater's edge. Since the device could record a spectrum every second, I could amass tens of thousands of measurements in a few hours. This enabled me to track each exhalation of the volcano. There were no samples to take back to the lab for analysis – I could process the infrared spectra on my laptop in the Hut during a spell of bad weather and have the results in a few hours.

Of course, in the field, things don't always go to plan, especially in Antarctica, where the environment is as inimical to electronics as it is to the body's extremities. I often had to take my gloves off to use screwdrivers or type instructions in -40°C (-40°F) of windchill, and my fingers would tingle for hours afterwards. The more or less daily effort to bring the spectrometer, automotive batteries, tripods, generators, scaffolding bars (on which to mount equipment) up and down the hill was backbreaking – imagine hauling the white goods in your kitchen up a sand dune in a double helping of winter clothing, and you get an idea of what it was like.

Once the spectrometer was running, it needed topping up with liquid nitrogen three times a day. I would rescue it if a storm was coming, but otherwise it would run autonomously for days. I couldn't know what the volcano would do, so relied on a continuous data record to capture unexpected events. In the words of an eighteenth-century engineer, François-Gabriel Courrejolles (who wrote a lost masterpiece on the triggering of eruptions and earthquakes): 'Nature often spontaneously unveils herself: but a long time is necessary to collect good observations.'[31]

Of the seven different volcanic gases I could measure, the most diagnostic were carbon dioxide, carbon monoxide and water vapour, since their proportions depended on the temperature and pressure of the magma. When I made a time plot of their quantities, an entirely unexpected repeating pattern emerged: the amount of carbon dioxide went up and down, while that of water vapour went down and up, roughly every ten minutes. It was so baffling, I thought it must be an artefact of my data processing – how could the different gas emissions vary so rhythmically? There was nothing to see in the crater that suggested anything other than a constant renewal of fresh lava at the surface of the lake.

While I could analyse an hour or so of measurements in the Hut, I couldn't process two or three weeks' worth of data – that required months of computing time. Then there were millions of datapoints to plot in dozens of different combinations, derivations of chemical conditions and temperatures, and mathematical analysis to quantify any trends. I could make a coffee in the time it took simply to open one of my spreadsheets. I had also recorded long sequences of heat images of the lava lake (technology had advanced enormously since my days spent pointing an infrared thermometer at Stromboli's vent). These showed how the magma was churning, and revealed the same ten-minute cycle – as if the lava were going on and off the boil. The gas composition and lake vigour were completely different characteristics measured using independent methods, so the pattern couldn't be spurious. I was amazed, even ecstatic – I had never seen anything like it before. It really did seem that Erebus had a beating heart.

Once home, I wrote up these findings for a scientific paper, but struggled to explain them.[32] You might wonder why that matters, and why anyone would get worked up about a volcano that is only ever threatening when scientists approach it. For me, the answer is that a repeating signal like this implicates underlying causes, the fundamental driving processes that lead a volcano to erupt in the way it does, and when it does, just as Frank Perret thought. If we can't explain such oscillations, or why the cycle repeats every ten minutes rather than every hour, there are basics of how volcanoes work that we are missing, undermining our prospects of forecasting volcanic activity. Looked at the other way around, though, if we *can* explain the behaviour, then we can develop better predictive tools.

On my last field mission to Erebus, in 2016, still trying to tackle the root causes of the cycles, we took a bespoke radar device. It was designed and built by a colleague, and the electronics fitted in a small case that was wired up to two large satellite dishes. These transmitted and received pulses of microwave energy and by aiming them at the lake we could measure the lava level. We found that it heaved up and down by a metre or two every ten minutes, as if the magma were breathing in and out.[33] It was the same cycle! I have never had a stronger sense of a volcano being an organic system.

Despite dozens of missions to Erebus by Phil and his colleagues over three decades, no one had ever spotted this behaviour – from the crater rim, the lava surface is just too far away to discern the pulsing by eye. It took sophisticated processing of data obtained with care using innovative equipment to reveal the rhythm of Erebus. And it took some more years to figure out what *is* behind the repeated and constant inhalation and exhalation of the volcano.[34]

It seems the key to Erebus' rhythm is an intermittent switch in the direction of magma flow between the lava lake and its underlying feeder pipe. As the bubbly molten rock rises up to the crater, the lava lake fills and gases fizz into the air. But all the while, the pressure at the bottom of the lake is rising because of the increasing weight of lava above – at a critical point, the flow reverses and lava drains out of the lake, back down the conduit supplying it. Now, the pressure below builds up, eventually forcing lava *up* again, and so the cycle repeats.

Thinking more about this explanation, it dawned on me that Professor David, in recognising a cadence to the sounds coming from the crater more than a century ago, was probably hearing these very reversals: when the lava was louder, it was rising, with bubbles sputtering; when it was quiet, the magma was sinking. More remarkable still, Shackleton's scientists seem even to have spotted the phenomenon from Cape Royds at the foot of the mountain. They took weather observations every two hours, and whenever it was clear, they looked at the Erebus smoke plume, since it served as a high-altitude weathervane. One winter's night, Shackleton recorded seeing the glow from the crater 'waxing and waning at intervals of a quarter of an hour through the night'.

Two of Shackleton's party, Professor David and the geologist Raymond Priestley, also drew a comparison between Erebus and Stromboli, noting that both erupted more often when atmospheric pressure was low. It's an enigmatic observation that I don't have an explanation for, but it perfectly demonstrates the value of observing and recording data. You never know what could be the key piece of information that slots into place, telling you something entirely new and unimagined. There are less-contested resemblances between the Eolian volcano and Erebus, though: they are similar in size (recalling that Stromboli rises from the deep seabed), and Erebus is also prone to strombolian eruptions, which result from the violent rupture of large gas bubbles. In fact, Erebus is the only volcano where I have actually caught a lava bomb – admittedly a small one, and it was tumbling down the side of the cone rather than falling through the air (and, having learned from my first experience with projectiles on Stromboli, I had gloves on).

On Erebus, these explosions can be very dramatic, with bubbles the size of the cupola of St Paul's Cathedral in London distending the lake, then popping dramatically, flinging huge dollops of lava over the crater rim. When they thud to the ground, they look like the turds of giants. They have a stunning golden lustre for a few days before the glassy strands that coat them start to corrode in the acid plume, spilling their cargo of feldspar crystals. Strangely, they are hollow, like tennis balls. Break one open, and the interior, with its bubbly texture and sinews of glass, resembles a human heart.

Conveniently, these explosions provide us with fresh lava samples that we can bring to the lab for analysis.[35] Like Lazzaro Spallanzani who carried out experiments on his rocks from Stromboli, we have pulverised and pressure-cooked our specimens to recreate the conditions the magma experienced below ground.[36] They've been further subjected to high-energy X-ray beams capable of nanoscale chemical measurements.[37] All this work has helped us to reveal the volcano's secret life below ground. For instance, we have traced the origins of carbon dioxide emissions from the lava lake to beneath the Earth's crust – in the mantle, as much as 15 miles down.[38] Given how carbon dioxide in the atmosphere regulates climate, it is vital to understand the volcanic source of the gas. What's more, we've shown how carbon dioxide drives profound chemical and physical changes in the magma it permeates, acting in a way to control the most fundamental traits of the volcano – it's size, shape and eruptive displays.

We also analysed the feldspar crystals, and found that the largest took decades to form. Like tree trunks thickening with each summer's ring, the crystals had grown as a result of multiple transits between deep magma reservoir and lava lake.[39] Such magma recycling had been invoked to explain the puzzling behaviour of other volcanoes, including Stromboli and Masaya, that belt out gas and heat, but not much lava. Now we had shown it really happens.[40] Although we had travelled to the end of the world to a volcano as remote from humankind as seems possible, we had proven a theory of general importance.

★

I think often of the 'heroic era' scientist-explorers who lived on Ross Island, surveying its terrain and fauna, sampling its rocks, measuring the air and tracking the weather. It was not so long ago – my grandparents' generation. The two world wars that followed claimed the lives of some of those men, and delayed lab work on the samples and collation of results – it was decades before all of the findings were published.[41]

Four years after Professor David's team reached the crater of Erebus, it was Raymond Priestley's turn to summit. He had been part of the original group that arrived in Antarctica with Shackleton, but had missed the chance to climb the volcano. Now twenty-six, he was a geologist on Captain Scott's *Terra Nova* expedition. He led a team of six, including another young protégé of David's, Frank Debenham (Deb), on the second ascent of Erebus.

On 2 December 1912 (the year the Hawaiian Volcano Observatory was founded), Priestley and Debenham set off with Tryggve Gran, a Norwegian ski specialist, Frederick Hooper, formerly a steward on the *Terra Nova*, and two navy men, Harry Dickason and George Abbott. They followed a similar route to David's at first, but then diverted to skirt up the side of the Fang glacier, close to where we pitch our acclimatisation camp. After ten days' toil, on the 12th day of the 12th month of 1912, they reached the crater.[42] Peering in, they may have been disappointed, for the volcano's depths were veiled by steam clouds. Unbeknownst to them, the team left traces that I would stumble upon, one hundred years later to the day.

As well as being fascinated by previous Erebus visitors (it is, after all, an exceptionally small club), my career had crossed over with Priestly and Debenham in other ways. After their work in the field, both moved to Cambridge to continue their studies and work on the wealth of data from Antarctica. Debenham helped to found the Scott Polar Research Institute in my university in 1920, and was appointed Cambridge's first Professor of Geography in 1931. So when I was back in Antarctica in 2012 I found myself scrolling through photographs of the ascent in the Scott Polar Research Institute's online archive. In one, entitled 'Highest camp in Antarctica', the men are seen erecting their tipi-style tent in front of a ridge of lava. Wouldn't it be a thrill, I thought to myself, to find exactly where they had been?

Not long after, I was firing up a snowmobile, with the image of the site fixed in my mind. It was a needle in a haystack mission and I doubted there would still be signs of their presence given the severity of Antarctic blizzards. But I was excited all the same and set off towards the old caldera wall, a section of which seemed to be in the background of the photograph.

The white ground was so brilliant and lacking contrast that it was hard to discern the gradient ahead. It would have been easy enough to nosedive into an icy ravine. After spotting a secluded valley in between two rocky crests, my gaze fixed on a crag with a diagonal fracture – it looked very familiar. I checked the photograph on the laptop, and scrutinised each nook and cranny and serration: a tower-like projection to the right, the shapes of tumbled boulders embedded in the snow. This was surely the site! I switched off the engine, stepped off the snowmobile and took a dozen paces closer. Then I noticed a semi-circle of rocks poking above the snow. Their orientation was too regular to be a chance scatter of eroded boulders from the low cliff. And there were three or four other rocks that fitted into the imaginary circle. They had to have been used to secure the flaps of the men's tent. Tingles ran up my spine.

I saw, in my mind's eye, Priestley, Abbot, Hooper and Gran fussing around their tent, congratulating each other for making it to the top. I couldn't help smiling, and instinctively called out, 'Hello boys!' I had the most intense feeling of reaching across the void and touching the past.[43] These men stepped on a truly pristine continent. They made the first maps, collected the first rock samples, and they had to find their way across treacherous sea ice, glaciers and mountains without any knowledge of what lay ahead. Our own work here is completely grounded in their legacy forged from courage, curiosity and unimaginable endurance.

*

I have listened attentively to Erebus, and would not like to guess how many days, months and years I have spent analysing and plotting and puzzling over the data we have collected there. Of all the volcanoes I have spent time with, Erebus has been the most persuasive in showing the importance of careful observations in underpinning new knowledge of the Earth. One hundred years ago, Debenham wrote that polar science is critical to understanding 'a globe of infinite diversity of form, climate, and process, all parts of which are interrelated'.[44] Having followed in his footsteps, I know exactly what he meant.

'Hypothesis-driven science' is a phrase that often appears in the academic world. When applying for grants for research, funding bodies want to know what you're setting out to prove. And this is why I always found it difficult to write proposals for our work on Erebus. Phil once joked to me, 'Our hypothesis is that Erebus is a volcano.' Erebus is an ever-changing yet reliable laboratory – each year we would innovate, bring new sensors and refine our questions and aims based on what we had learned the previous season.[45] We also brought and supported students and post-docs, artists, writers and filmmakers. I truly believe we did some amazing science and outreach work, but all this is impossible to capture in a hypothesis. Ross, Shackleton and Scott did not go to Antarctica to evaluate a proposition – they went there for the prospect, joy and significance of discovery.

On my first expeditions to Erebus, I merely eavesdropped on the volcanic activity. On later trips, I went there to listen, as carefully as I could, to the volcano's vibrations. I had to be ready in a moment to adapt my strategy according to the wind, humidity, temperature and magmatic menace. In this way, the measurements revealed many striking – and some still-perplexing – phenomena. I learned from Erebus what Frank Perret had discovered a century before: the inner Earth has a pulse.

At Erebus, I have come closer to perceiving the arcane mechanisms and mysteries of a volcano than anywhere else I have worked. And I increasingly came to understand how so much of what we do follows in the footsteps of pioneers. One of them, an unnamed officer aboard HMS *Erebus,* wrote the following 180 years ago:

Nowhere is it more apparent that Nature herself is the greatest of artists than in the polar regions. And it is there, confronting the limits of human endurance in seeking to penetrate the mysteries of Nature, that the art of observation, experimentation and theorising itself becomes Sublime and magnificent.

The forerunners of Antarctic exploration learned things beyond their imaginations – and they did so by probing the unknown, without hypotheses, without preconceived notions or foregone conclusions, by bringing together people from different fields.

The end of each field season on Erebus was, for me, marked by a wrenching pang of loss as we were lifted into the bright sky aboard the helicopter, and effortlessly banked round the side of Erebus, a spectral cloth of cloud spreading in its lee, sun glint blazing the icy sea. Within fifteen minutes, we would be back in McMurdo; in twenty-four hours, New Zealand; in three days, home. Each time, I felt the heartache of leaving a lover I might never set eyes on again. Even now, years since my last Erebus mission, whenever I see a piercing low sun on a crisp winter's morning my thoughts turn again to the mountain and the wonder of being there. I think Shackleton must have felt the same for he wrote this in a poem titled 'Erebus'[46]:

> *With icefield, cape, and mountain height, flame rose in a sea of gold.*
> *Oh! Herald of returning Suns to the waiting lands below;*
> *Beacon to their home-seeking feet, far across the Southern snow.*
> *In the Northland in the years to be, pale Winter's first white sign*
> *Will turn again their thoughts to thee, and the glamour that is thine.*

The Volcano and You

Breakfast at the crater.[1]

*'I am 3½ years old. I really like studying volcanoes with my
parents. I want to know what it feels like to see a volcano.'*
— Lucy, Rogersville, Tennessee (2020)[2]

If you are one of the billion or so people settled within fifty miles of an active volcano – defined as one that has erupted within the past 12,000 years – then you may already know its aura.[3] In Seattle, Quito, Naples, Tokyo, Auckland and many other cities lying close to latent magma, the urban skyline is embellished with one or more cratered peaks. Such volcanoes are invariably entwined with the history and culture of the region – what would Naples be without Vesuvius, Seattle without Mount Rainier, or Tokyo without Fuji?

But for most people, encounters with volcanoes are rare and represent journeys into realms where nature feels at once primal, temporary and fragile. There seems to be something almost universally attractive about volcanoes; a magnetism that speaks to an ancestral enchantment – like the feeling we get seeing colour gleam through stained-glass windows. Children often have a volcano phase, repeatedly drawing the satisfying cone shape with sparks bursting out, and for many of us that curiosity never goes away. For those who still feel the pull and are ready to immerse themselves in these landscapes, there have never been better opportunities – the title of this chapter is an overture not an omen.

The human urge to experience volcanoes goes way back – one of the earliest well-documented examples of volcano tourism dates from the 2nd century BCE when well-heeled Romans began thronging the seaside resort of Baiæ, situated within the Campi Flegrei caldera near Naples. Sulphurous thermal pools were the main attraction – bathing in them was reputed to cure ailments and rejuvenate the body.[4] Recreational soaks were still popular in Sir William Hamilton's time – one of the illustrations in his *Campi Phlegræi* shows a bather at Solfatara. But by the British envoy's time, the Grand Tour was in full swing, the flaming Vesuvius its climax. Some, like the chemists Sir Humphrey Davy and Robert Bunsen, brought their portable laboratories to study the gushing gas and lava but most came for the frisson of a perilous encounter with the untamed fire mountain. It was arguably the beginning of mass geotourism.

For those unwilling or unfit to climb, mules could be hired for part of the journey, and for the last steep stretch on unbound cinders, visitors could pay to be hauled up with the aid of a rope tied around the waist. Breakfast or lunch would be prepared over hot crevices beside the crater 'and eaten with a greater relish in the very jaws of possible destruction', as one correspondent put it.[5]

Today many volcanoes are 'drive-ins' with catering facilities, obviating the need to be dragged up the mountain or cook your dinner in a fumarole. Among the most iconic are a super-volcano that also has the distinction of being the world's first national park: Yellowstone, formally protected as a 'pleasuring-ground for the benefit and enjoyment of the people' on 1 March 1872.[6] Alternatively, if you are looking for something at the other end of the explosivity scale then the Hawai'i Volcanoes National Park is for you. Both are immensely popular: Hawai'i welcomed 1.3 million visitors in 2021, Yellowstone an astonishing 4.9 million, according to the National Park Service.[7]

Yellowstone and Hawai'i Volcanoes National Park are both UNESCO World Heritage Sites, another eighty of which boast volcanic features.[8] UNESCO also maintains a list of Global Geoparks, an initiative that helps communities marry the often conflicting goals of developing tourism while protecting the environment. As well as their biological and geological significance, volcano geoparks might celebrate mining heritage, or commemorate catastrophe.[9] They include the Reykjanes peninsula of Iceland, Toba caldera in Indonesia and Ngorongoro in Tanzania.[10] In total there are nearly 200 geoparks, spread across nearly fifty countries – all invite you to unfetter your feet.[11]

If you don't reside near a sleeping giant, extinct volcanic terrain may yet be close at hand. The British Isles, for example, abound in igneous attractions, including the Giant's Causeway in Northern Ireland, Snowdon in Wales, Glencoe in Scotland, Scafell in the Lake District and Arthur's Seat in the centre of Edinburgh. While these landmarks might lack Mount Fuji's stature and symmetry, they invite deep geological time travels, and with the help of interpretative guides their former pyroclastic vigour can be imagined. At an altogether grander scale, the Deccan Traps in western and central India span an area the size of Spain. This huge volume of basalt lava, erupted at the end of the Cretaceous Period, epitomizes the risk and reward of volcanism – it spelt the end for the dinosaurs but opportunity for mammals.[12] It is fitting that temples hewn into the lava 1,500 years ago were dedicated to Shiva, the god of destruction and regeneration.[13]

With their disarrayed tracts of lava, pumice and slag, many might consider volcanoes barren ground. In reality they form ecosystems rich in flora and fauna adapted to their singular habitats, such as Darwin's finches on the Galápagos islands, the spectacular red-flowering buglosses of Teide volcano on Tenerife and the Japanese macaques famed for 'taking the waters' in geothermal pools.[14] Even some of the most severe volcanic environments such as the ice caves of Erebus or the hot acid crater lake of Ijen on Java teem with extremophile microbes, some of which may one day find use in medicine, or for the clean-up of pollutants.[15]

In wet and warm climates, volcanic deposits quickly degrade to mineral-rich soils that nourish rampant cloud forests and bountiful crops.[16] This is one reason why Central Java in Indonesia, home to Merapi volcano and its subterranean kingdom, is among the most densely populated regions of the world. So even if home for you is far from the volcanic belts, it is likely some of the grains, fruits and nuts, wines, pulses and coffee in your kitchen cupboards were nourished by volcanic soil. Consuming these foods enmeshes you with the great geochemical cycles of carbon, water and nutrients that interlace with the Earth's geophysical pulse of eruptions, rifting and subduction.[17]

As well as being rich in agricultural terms, volcanoes have long been fertile ground for spiritual life. The German botanist and volcanologist Franz Junghuhn, wrote of Java that 'Nothing reveals the harmony of creation so clearly as a journey from the high volcanic interior of a tropical country to the coast'.[18] He'd experienced the magic and menace of volcanoes close up but saw, too, how they cradle the soul of a place and its people; how they are gateways to other worlds; why the gods call the mountains home.

We give meaning to volcanoes through our attention, whether cultural, spiritual, aesthetic, extractive or scientific. For communities such as those of Java, so deeply connected with their volatile volcanic neighbours, culture is bound with landscape, memory with unfathomable time, cosmology with underworlds. In Campania in southern Italy, to look upon the alternating layers of pumice and archaeological materials at Herculaneum or Pompeii is not to stare into the past but into the mirror.

The entanglements of people and place, of nature and culture, run deep at Yellowstone and on Hawai'i, too. The super-volcano's prolific obsidian outcrops were utilised by Native Americans for more than 11,000 years,[19] while Native Hawaiians regard

Kīlauea and Mauna Loa volcanoes as their own kith and kin.[20] It is such genealogy that requires conservationists and park services to work together with ancestral custodians of the land, to ensure simultaneous protection of biodiversity, 'geodiversity' and traditions.[21]

'Geoheritage' has a growing status in volcanology. As a focal point for conversations and collaborations spanning the natural and social sciences, the arts and humanities, it's a reminder that anyone can get involved in the study of volcanoes. This was true for many extraordinary people who had no formal training in geology yet who sought to understand the positions of volcanoes on the globe and to probe their infernal realms, often at Promethean risk. Among them were priests, doctors, chemists, biologists, explorers, soldiers and, not least, a college dropout who caught the volcano bug in his mid-thirties, Frank Perret.

Like Junghuhn before him, Perret was a visionary whose gaze went beyond science: for him, volcanoes made the world fit for human purpose. 'It is in the operations of its persistent volcanism' he wrote, 'that we can best observe the adjustments of the living earth in its progress toward a perfect habitation for man'.[22] Even if we don't fully subscribe to this view, there is no doubt that volcanoes mean more to us that disaster, doom and destruction. In their long and varied lives they have actually sustained humanity, shaped the evolutionary paths we have taken and inspired us to ask bigger questions about our place in the universe. Should our descendants colonise a forbidding Mars with the help of 4th Planet Logistics (whose CEO I met in Iceland), they might be hunkered down in lava tubes, breathing oxygen extracted from basaltic rock.[23] It might not be long before they are venerating the spirits of lofty Olympus Mons.

Earthlings becoming Martians might be one thing, but could life start on a planet without volcanoes? There are several reasons to think not – on Earth, at least, volcanoes seem to have been essential to the origins of life. Among the first to wonder how chemistry might have become biology was none other than Charles Darwin. He wrote to his friend, Joseph Hooker (thirty years after the botanist had laid eyes on Erebus volcano with Captain Ross): 'if (& oh what a big if) we could conceive in some warm little pond with all sorts of ammonia & phosphoric salts,—light, heat, electricity &c present, that a protein compound was chemically formed, ready to undergo still more complex changes . . .'.[24]

It was in the 1950s that the American geochemist Stanley Miller actually synthesised possible ingredients of Darwin's primeval pool in the laboratory. By injecting water vapour, carbon monoxide, ammonia, and hydrogen sulphide into a glass flask and discharging a 60,000 volt electrical spark across the gas mixture, Miller manufactured numerous complex organic molecules including amino acids, the foundations of proteins.[25] What makes this experiment so interesting is that spectacular lightning often scintillates swelling ash clouds, as Sir William Hamilton recorded after witnessing an eruption of Vesuvius in 1779:

> the black smoke and ashes issuing continually from so many new mouths, or craters, formed an enormous and dense body of clouds over the whole mountain, and which began to give signs of being replete with the electric fluid, by exhibiting flashes . . . of zig-zag lightning . . .[26]

This is one way that the organic precursors of life could have originated but they would still have needed to interact in a hot, watery environment such as we find today in the geysers and hot springs of Yellowstone, or perhaps the pores of floating pumice,[27] in which molecules could interact, complexify and take on functionality.[28] Other scientists trace the origins of life to the dense grey-black plumes of hot, mineralised water that jet from vents along the volcanically active ocean ridges. Almost as soon as they were discovered in the 1970s, there was speculation that these 'hydrothermal vents' represented factories for life's building blocks.[29]

Whether or not we can trace our evolutionary roots back to it, volcanism certainly helped to sustain creation and drive it forwards. The American volcanologist Thomas Jaggar wrote that 'Man is largely a puff of hydrogen'. In fact, humans are made up of around 10% hydrogen by mass, but Jaggar's conflation of our bodily composition and the breath of volcanoes is not so wide of the mark. Our other main constituents are oxygen (65%) and carbon (18%). These proportions are quite similar to those found in volcanic gases such as blow from the lava lake of Mount Erebus. The similarity is not entirely surprising – both the body and volcanic gases are water-rich – but it goes deeper than this. Although we are ultimately made of 'star stuff' as Carl Sagan explained, that stuff ended up coalescing into planets, and on our mobile Earth it was kept in a state of flux through tectonic and volcanic action. The emissions from our mountains of fire regulated climate and charged the Earth's water towers to make the world habitable for billions of years.

In an Anthropocene world that seems increasingly out of kilter, volcanoes offer opportunities beyond geotourism. Their resources have attracted us since our ancestors fashioned the first obsidian tools in the shadows of volcanoes of the East African Rift Valley. Today it is their heat that we tap. Geothermal energy yields significant fractions of national supply in several volcanically endowed countries, including Kenya, Iceland, the Philippines and New Zealand. Volcanic terrain could even help to rebalance the carbon cycle. One idea is to capture carbon dioxide and store it in basalt, the most abundant rock on Earth.

As well as oxygen, basalt is rich in calcium, magnesium and iron, all of which react with carbon dioxide to make carbonate minerals – up to 70kg of the greenhouse gas could realistically be stored in a cubic metre of basaltic rock.[30] It might sound unfeasible to accelerate the rock weathering cycle in this way, but a plant in Iceland has already been doing it for more than a decade.[31] Given the extensive basalt terrains of the Earth, such as the Deccan Traps in India, if the process can be replicated at scale it could help to alleviate global warming.[32]

Volcanoes, then, are not simply sites of disasters or of spectacular beauty or purveyors of awe-inspiring pyrotechnics. They are places that connect past, present and future; where land, water and sky animate custom, belief and knowledge, and vice versa. They helped fashion the human condition – our spirit, beliefs, respect for nature, resilience, adaptability and fear. Those shamans, explorers and scientists who were motivated to observe and study volcanoes across the world made their discoveries and adventures part of their own stories – and their legacy of knowledge helps to anchor the whole human narrative. Through the climate change wrought by the largest eruptions, our world history is entangled with volcanoes more than we realise. We may all feel the repercussions of the next really big event. Seeing the world through this lens gives us new perspectives on agency, causation and correlation.

Volcanoes make us aware, not just to feel and sympathise with the story of humankind, but to draw us all into the bigger mysteries of the soul and the cosmos. They remind us we're not just bodies hurtling through the universe – we're something beyond what we know. This doesn't make volcanoes less dangerous to live alongside, but having a greater understanding does make them less strange.

Volcanoes made us what we are, and long after we are gone they will hold the memory of our existence.

Acknowledgements

I received an email in April 2020 asking if I'd consider writing a book on volcanoes. I might have politely declined had not Anna Baty at Hodder and Stoughton elaborated that she had in mind 'a tapestry of science and culture and adventure that shows us how the science interacts with the social and what it means to live on a planet that is enthral to these powerful forces . . .'. I sensed at once a shared fascination for the multiple meanings of volcanoes and the diversity of knowledge systems by which they are understood.

I received brilliant feedback, counsel and enthusiasm on the text at many stages from Anna, as well as from her colleagues Izzy Everington, Joe Calamia (at University of Chicago Press) and Zakirah Alam. Tara O'Sullivan expertly copyedited the manuscript, and Holly Ovenden designed the magmatastic cover. I thank them and all those behind the scenes at the publishing houses.

I am very grateful to Valeria Amoretti, John Binns, Simon Schaffer, Jenni Barclay and Ross Mitchell for their comments and suggestions on the draft manuscript, and to Kosima Weber-Liu, Lamya Khalidi, Nicola Di Cosmo, Rich Stone, Willy Aspinall and James Hammond for guidance on specific points.

Volcanoes inspire me, of course, but so do my colleagues, none more so over the past years than my friend Ulf Büntgen. Among many initiatives he involved me with is the disciplinary melting pot he curated at the Centre for Interdisciplinary Research (Zentrum für interdisziplinäre Forschung, ZiF) at the University of Bielefeld. I am grateful to the staff at ZiF and all the meeting participants, whose manifold perspectives suffuse this book.

One of the joys of being a volcanologist is belonging to a worldwide community, and the reward of being an educator is seeing students and early career researchers flourish. There are so many I am grateful to for their collaboration and inspiration spanning the many years since Peter Francis and David Rothery set me off on my journey. Here (in addition to the aforementioned) I name those I've worked most closely with: Amy Donovan, Andy Woods, Anja Schmidt, Bill Rose, Bruno Scaillet, Céline Vidal, Christine Lane, David Pyle, Evgenia Ilyinskaya, Gezahegn Yirgu, Giuseppe Salerno, Jan Esper, Letizia Spampinato, Lori Glaze, Marie Edmonds, Markus Stoffel, Mike Burton, Nial Peters, Patrick Allard, Phil Kyle, Philipson Bani, Pierre Delmelle, Séb Guillet, Seife Berhe, Stefan Kröpelin, Tamsin Mather, and Yves Moussallam.

From the humanities and other branches of sciences, the following have all inspired me: Anna Guðjónsdóttir, Anil Seth, Armin Linke, George Finlay Ramsay, Ilana Halperin, Jane Munro, Jemila MacEwan, Karen Holmberg, Lena and Werner Herzog, Malena Szlam, Michael Bravo, Michael Madsen, Nana Pohjanpalo, Peter Baxter, Peter and Silvia Zeitlinger, Philip Ursprung, Sara Dosa, Terry Glaze, Tim White and Tom Adams. I also thank Agnès Berthin-Scaillet, Alice and Roger Barford, Anna and Paul Krusic, Anne and Michael Herzog, Catherine and Nico Leroy, Chief Mael Moses, Christiane Büntgen, Franc and Bill Adams, Maria and Ben Goodall, Jamie Coello Bravo, Ri Kum Ran, and Shimrit and Over Ziv for friendship and collaboration.

My research and filmmaking has been funded by various organisations but for more recent support I am extremely grateful to the Leverhulme Trust, the National Science Foundation's U.S. Antarctic Program, the European Union, the Simons

Foundation, the Diamond Light Source and the UK Natural Environment Research Council.

At home, I thank Anna Barford for enduring love, and Poppy and Maya for ensuring the writing wasn't rushed.

I dedicate *Mountains of Fire* to my colleagues, old and new, and to all the volcanologists to come.

Notes

Epigraph

1 R. Rogaski, 'Knowing a Sentient Mountain: Space, science, and the sacred in ascents of Mount Paektu/Changbai', *Modern Asian Studies* 52 (2018): 716–752.

Dreamland of the Living Earth

1 J. Ogilby, '*America: being the latest and most accurate description of the New World*', (London, 1671).
2 J. Oliver, 'Earthquake seismology in the plate tectonics revolution', in N. Oreskes (ed) *Plate Tectonics: An insider's history of the modern theory of the Earth* (Boulder, Westview: 2003), 155–166.
3 T. A. Jaggar, *My Experiments with Volcanoes* (Honolulu: Hawaiian Volcano Research Association, 1956). One year before the Hawaiian Volcano Observatory opened (the first in the United States), Japan's first volcano observatory was inaugurated on Mt Asama – at 2,000 metres (more than a mile) elevation. Being close to the crater it was snowed in over winter and within range of lava projectiles.
4 A great example of phenomena we can record at the surface, helping us to visualise the unseen world beneath, is how observations of the seabed, of earthquake distributions, and of the magnetism and age of rocks led to the 'plate tectonics revolution' in the 1960s. Most volcanic and seismic activity is associated with the motion of tectonic plates, which is driven by deeper flows of heat and rock. The new understanding of our dynamic

planet solved many puzzles, including the long-standing conundrum of why the coastlines of Africa and South America look like adjoining puzzle pieces – they had been part of a larger continent until the Cretaceous period, when they split into individual plates that have separated ever since at about the same rate at which fingernails grow. For representations of global plate models, see: D. Hasterok *et al.*, 'New maps of global geological provinces and tectonic plates', *Earth-Science Reviews* 231 (2022): 104069. For more on what plates get up to, see R. Mitchell, *The Next Supercontinent: solving the puzzle of a future Pangea* (Chicago: University of Chicago Press, 2023).

5   J. Carrillo, 'From Mt Ventoux to Mt Masaya: The rise and fall of subjectivity in early modern travel', in *Voyages and Visions: Towards a Cultural History of Travel*, eds. J. Elsner and J. P. Rubiés (London: Reaktion Books, 1999): 57–73.

6   D. Turner, 'Forgotten treasure from the Indies: the illustrations and drawings of Fernández de Oviedo', *The Huntington Library Quarterly* 48 (1985): 1–46.

7   Among other wonders, Oviedo's drawing of a pineapple was the first picture of the fruit to be published. On Oviedo's contributions, see: A. C. de la Rosa, 'Representing the New World's nature: Wonder and exoticism in Gonzalo Fernández de Oviedo y Valdes', *Historical Reflections/Réflexions Historiques* (2002): 73–92.

8   One of his senior officers, Diego de Ordás, begged to tackle the snowy summit. Cortés consented, bidding him to take two other soldiers for company. Local guides led the group halfway up the volcano, to where a cluster of temples had been constructed, but they refused to go further, explaining that the spirits of tyrants dwelt there. Undeterred, the Spaniards continued, but had not got far when the ground shook violently and 'huge flames of fire, half burnt and perforated stones, with a quantity of ashes' shot from the peak. After an hour, the outburst subsided and the men resumed the ascent. Eventually, they reached the fuming

crater. Peering in, they spied in its depths a fire that seethed like a furnace of molten glass. Away from the smoke, the air was crystal clear, and Ordás admired the sweeping view of the Valley of México, which encompassed Tenochtitlan, the Aztec capital, on an island connected by bridges to the shore of Lake Texoco. Ordás knew the view constituted valuable military intelligence and committed it to memory. When they could bear the fumes, thin air and cold no more, the men retraced their steps. Their turnaround was timely, as another belch from the crater assaulted them with 'boulders of burning fire', but they found protection below a protruding rock. The men's return was met with astonishment and celebration. On his return to Spain, Ordás was granted a coat of arms featuring a smoking volcano. (I want one, too.) Considering the volcano's altitude (well over 5,000 metres) and vertiginous snowy slopes, for which the climbers were ill-equipped (they wore sandals and used swords as ice axes), it is remarkable they summited. The epic adventure is the focus of the 2015 film *Epitafio*.

See: Bernal Diaz del Castillo, *The Memoirs of the Conquistador Bernal Diaz del Castillo*, Vol 1 (of 2), trans. J. Ingram Lockhart (London: Hatchard, 1844); F. Cervantes de Salazar, *Crónica de la Nueva España*, (Madrid: The Hispanic Society of America, 1914), Book 3, Chapter 58; Ismael Arturo Montero-García, 'Arqueología e historia de los volcanes Popocatépetl e Iztaccíhuatl, México', *Revista de Arqueología Americana* 34 (2016): 187–222.

9 The Spaniards quickly reached Tenochtitlan, where they infiltrated the Aztec palace and took Moctezuma hostage. But Cortés was running low on gunpowder, without which he could neither defend himself nor conquer more of the New World. Of its ingredients, charcoal was readily obtained, and saltpetre available in bat caves. Cortés recalled seeing sulphur atop Pico de Teide volcano on Tenerife, and his mind turned to Popocatépetl. He summoned his gunner, Montaño, and another

soldier, Mesa, inciting them: 'It is your destiny to preserve what we have gained and to acquire great kingdoms and lordships!' Montaño replied: 'We will bring back the sulphur or die trying.'

He chose three further companions, and the party had slogged part way up the volcano by nightfall the next day. At sunrise, they resumed the battle. There was no clear pinnacle to aim for, just an interminable trudge, each footfall on yielding cinders or slick ice. After several hours, suffering from altitude sickness and snow blindness, they reached the summit. Peering into the crater, below cliffs striped with sulphur deposits, they spied a fierce glow 'like a natural fire'. They scouted for a way to get at the precious seams, and drew lots to resolve who would enter first: Montaño! He made seven descents before handing over, advising, 'Whatever you do, don't look down!' Their baskets full, all that remained was to get back safely, and by dawn they were marching into the Spanish headquarters at Cuyoacán. The sulphur was refined and furnished fifty kegs of gunpowder, enough firepower to continue subjugating the indigenous peoples of Mesoamerica.

The Spanish chronicler Cervantes considered the conquest owed much to the courage of Montaño, who, for his part, vowed never to return to the crater, 'not even for all the treasure in the world'. It seems no one set foot at the summit again until 1827, a few years after México's independence. This ascent took four days of toil. It rekindled interest in the sulphur deposits, and between the 1850s and 1870s a modest extractive industry resumed. The miners, *volcaneros*, descended into the crater using a windlass, and piled rocks around gas vents to condense sulphur vapours. It was terrible work at altitude with a constant threat of rockfalls. Many died riding the sacks filled with sulphur like sleds down the icy flanks of the volcano.

See: Cervantes de Salazar, *Crónica de la Nueva España*, Book 6, Chapters 7–11; J. Norman, 'Sulphur from Popocatépetl',

Journal of Chemical Education 11 (1934); and M. C. LaFevor, 'Sulphur mining on Mexico's Popocatépetl Volcano (1820–1920): origins, development, and human-environmental challenges', *Journal of Latin American Geography* (2012): 79–98.

10 Vulcano is one of several volcanoes beneath which, it was said, Zeus's weapons were forged by Cyclopes. Thank goodness we didn't end up with 'Popocatépetl' or 'Eyjafjallajökull' as the generic term for mountains of fire. See: P. Aebischer, 'Esp. *volcan*, it. *vulcano*, fr. *volcan*: une conséquence de la découverte de l'Amérique centrale', *Zeitschrift für Romanische Philologie* 67 (1951): 299–318; K. L. Taylor, 'Before volcanoes became ordinary', *Geological Society, London, Special Publications* 442 (2017): 117–126.

11 Translated in: J. G. Viramonte and J. Incer-Barquero, 'Masaya, the "Mouth of Hell", Nicaragua: Volcanological interpretation of the myths, legends and anecdotes', *Journal of Volcanology and Geothermal Research* 176 (2008): 419–426.

12 Oviedo had settled in León, the old capital of Nicaragua, by the late 1520s. The account of his expedition and commentary on Blás de Castillo is from: Fernández de Oviedo and Gonzalo Valdés, *Historia general y natural de las Indias, islas y tierra-firme del mar océano*, Vol. IV (Madrid: Imprenta de la Real Academia de la Historia, 1855), Book 42, Chapters 5–10. Translations into English of parts of Oviedo's account of Masaya can be found in: E. G. Squier, *Nicaragua: Its People, Scenery, Monuments, and the Proposed Interoceanic Canal: in two volumes* (London: Longman, Brown, Green and Longmans, 1852). A project is underway to translate the entire *Historia* into English for the first time – search for 'The Oviedo project at Vassar' online. The resemblance of the name Masaya to 'Messiah' is odd given the history of European attitudes to the volcano, but coincidental. Oviedo tells us the Chorotegans called the volcano *Popogatepe* and that it meant 'mountain that boils'. He also says 'Massaya' [sic] means 'burning mountain'. The etymology seems uncertain,

though. One theory is that it comes from *mazatl-yan* – 'place of deer' – and another that its linguistic roots are in México and that it signifies 'where it rains fire'. See: J. Incer, *Toponimias indígenas de Nicaragua* (San José: Libro Libre, 1985).

13   Now there is a sealed road that winds up to the crater, but I have made the mistake in the past, when walking up, of attempting 'shortcuts' across the blocky lavas. It is forbidding terrain that could offer the ultimate challenge for an endurance race.

14   Apologies, I couldn't find any online calculator to convert these figures to imperial or metric units. Oviedo made the first map of the crater region, showing the lava pit within nested craters.

15   Nicaragua's former President Somoza is said to have liquidated political opponents in the same way in the 1960s.

16   In Europe, thanks to Oviedo, Masaya became the best-known volcano in the Americas for a long time. George Poulett Scrope, in his 1825 'textbook' on volcanoes, likened the persistent activity of Masaya, which he called 'the Devil's Mouth', to that of Stromboli in Italy. G. P. Scrope, *Considerations on Volcanos* (London: Phillips, 1825).

The updated 1862 edition of Scrope's work has as good a definition of an 'eruption' as any to be found: 'the forcible expulsion of heated matters, gaseous, fluid, or solid (usually of all three together), from the interior of the earth upon its surface'.

17   A thriving colonial city, until ruined by earthquakes in 1610 and then entombed in ash spewed from nearby Momotombo volcano. León Viejo was inscribed in the World Heritage list in 2000 and uniquely preserves the early colonial Spanish architecture.

18   S. Winchester, *Krakatoa: the day the world exploded* (New York: Viking Press, 2003).

19   There is a great example of this in the outskirts of the Nicaraguan capital. It was brought to international attention by Dr Earl Flint. Though a physician, he spent much of his time procuring antiquities for Harvard University's Peabody Museum. In 1883,

just as his own home was being sprinkled with ash from the neighbouring volcano, he heard that footprints had been uncovered in a bed of 'volcanic detritus and ash' by working quarrymen. Flint attributed the tracks to 'pre-Adamite' humans living 200,000 years ago, and hailed them as 'the most important discovery touching man's antiquity'. Another American contemporary of Flint living in Nicaragua, geologist John Crawford, wrote 'the footprints indicate haste, confusion and excitement [as if seeking shelter] from a storm of burning hot volcanic ashes and cinders'. This notion is corroborated by more recent analysis, which attributes the eruption to Masaya volcano. But the prints are much younger than Flint reckoned, being closer to 2,000 years in age. The site, known as Las Huellas de Acahualinca, is open to the public and makes a worthwhile side-trip on any visit to Managua.

See: E. Flint, 'Human foot prints in Nicaragua', *The American Antiquarian and Oriental Journal* 6 (1884): 112; E. Flint, 'Human foot prints in Nicaragua', *The American Antiquarian and Oriental Journal* 7 (1885): 156; E. Flint, 'Pre-Adamite footprints', *The American Antiquarian and Oriental Journal* 8 (1886): 230; J. Crawford, 'Neolithic Man in Nicaragua', *The American Antiquarian and Oriental Journal* 13 (1891): 293; M. G. Lockley *et al.*, 'America's most famous human footprints: history, context and first description of mid-Holocene tracks from the shores of Lake Managua, Nicaragua', *Ichnos* 16 (2009): 55–69; W. Perez *et al.*, 'The Masaya Triple Layer: a 2,100 year old basaltic multi-episodic Plinian eruption from the Masaya Caldera Complex (Nicaragua)', *Journal of Volcanology and Geothermal Research* 179 (2009): 191–205; and H. U. Schmincke *et al.*, 'Walking through volcanic mud: the 2,100 year-old Acahualinca footprints (Nicaragua) II: the Acahualinca people, environmental conditions and motivation', *International Journal of Earth Sciences* 99 (2010): 279–292.

20   Take carbon dioxide, for example, which is formed from a carbon atom bonded on either side to an oxygen atom, like this: O=C=O. When infrared radiation falls on the gas, it absorbs precisely defined wavelengths of the light, causing the atoms to shift in relation to each other and warming up the molecule (hence carbon dioxide's 'greenhouse' effect). A spectrum of infrared light that has passed through CO_2 will reveal a complex pattern of dark lines and bands, that both identify the gas and quantify how much is present. This is how the 'Keeling curve' (https://keelingcurve.ucsd.edu/) – the historical time series of atmospheric measurements of CO_2 made at the Mauna Loa Observatory on the Big Island of Hawai'i – is obtained. Since 1958, when data collection began, atmospheric CO_2 abundance has risen by a third.

This increase has nothing to do with volcanoes. The CO_2 emitted by *all* the world's volcanoes, on land and in the oceans, is at most two per cent of that generated by fossil fuel use. The last time volcanoes and humans were on a par for CO_2 output was in the mid-nineteenth century. See: T. M. Gerlach, 'Present-day CO_2 emissions from volcanos', *Eos, Transactions American Geophysical Union* 72 (1991): 249–255. M. R. Burton *et al.*, 'Deep carbon emissions from volcanoes', *Reviews in Mineralogy and Geochemistry* 75 (2013): 323–354.

21   'Proclivity' shares its etymology with 'Clive', from the Old Norse *klîfa*, which means to climb. I ended up in the right profession! The word 'observation' also has interesting origins. In Latin *obseruatio* meant, among other things, the inspection of omens, as well as keeping a watchful eye on potential threats.

22   Of course, inequalities, inequities and injustices prevail, for instance in career progression and access to resources, and tensions still brew when pure and applied research priorities clash, in post-colonial contexts, and where local communities are threatened both by a volcano and enforced resettlement. For some perspectives on contemporary debates in volcanology, see:

J. L. Kavanagh *et al.*, 'Volcanologists—who are we and where are we going?' *Bulletin of Volcanology* 84 (2022): 53. J. Barclay *et al.*, 'Disaster aid? Mapping historical responses to volcanic eruptions from 1800–2000 in the English-speaking Eastern Caribbean: their role in creating vulnerabilities', *Disasters* 46 (2022): S10–S50. S. Saha *et al.*, 'A place-based virtual field trip resource that reflects understandings from multiple knowledge systems for volcano hazard education in Aotearoa NZ: Lessons from collaborations between Māori and non-Māori', *Journal of Geoscience Education* (2022): doi: 10.1080/10899995.2022.2109397. R. Christie *et al.*, 'Fearing the knock on the door: critical security studies insights into limited cooperation with disaster management regimes', *Journal of Applied Volcanology* 4 (2015): 19.

23 For a thoughtful discussion, see: F. Mazzocchi, 'Western science and traditional knowledge: Despite their variations, different forms of knowledge can learn from each other', *EMBO reports* 7 (2006): 463–466.

24 The eruption of Mount Mazama around 5700 BCE formed the deepest lake in the United States: Crater Lake in Oregon. It ranks among the largest eruptions of past millennia – similar in scale to that of Tambora in 1815. Ancestral knowledge of the eruption and its distant impacts has been preserved amongst Native American communities through sacred storytelling. See E. E. Clark, *Indian Legends of the Pacific Northwest* (Berkeley: University of California Press, 1953). G. A. Oetelaar, 'Natural disasters and interregional interactions: La Longue Durée in Northern Plains historical developments', *Plains Anthropologist* 66, no. 257 (2021): 1–33. See also: P. Nunn, *The Edge of Memory: Ancient stories, oral tradition and the post-glacial world* (London: Bloomsbury Publishing, 2018).

25 See, for example: S. Ramos Chocobar and M. Tironi, 'An inside sun: Lickanantay volcanology in the Salar de Atacama', *Frontiers in Earth Science* 10:909967 (2022).

Land of God

1 L. Spallanzani, '*Viaggi alle due Sicilie e in alcune parti dell'Appennino*', vol. 2, Plate 3 (Pavia: Baldassare Comini, 1792).

2 H. Tazieff, *Craters of Fire* (Hamish Hamilton: London, 1952): 185.

3 Just three weeks into the nearly four-month-long production, Bergman received a letter from the vice-president of the Motion Picture Association of America, Hollywood's moral police, warning that the affair would destroy her career, and urging her to deny the rumours she was leaving her husband and ten-year-old daughter. It's difficult to imagine how Bergman completed the film in the grip of such a profound personal crisis; perhaps she was able to channel her emotions into conveying the isolation and opprobrium Karin experiences.

4 Johnson called for federal censorship legislation to control the private behaviour of actors, directors and producers. The *Washington Star*'s critic typifies contemporary reviews of the film: 'a miserably inept work, meaningless as to story, grotesque in performance, confused in direction, and profoundly dull'. See: O. Gelley, 'Ingrid Bergman's star persona and the alien space of "Stromboli"', *Cinema Journal* 47 (2008): 26–51. Also: I. Bergman and A. Burgess, *Ingrid Bergman: My Story* (London: Michael Joseph, 1980). There is a good documentary feature telling the story of the making of *Stromboli* and a rival film, *Vulcano* (starring Rossellini's spurned lover, Anna Magnani): *La guerra dei vulcani*, directed by Francesco Patierno (2011).

5 M. Wyke, 'Mobilizing Pompeii for Italian Silent Cinema', *Classical Receptions Journal* 11 (2019): 453–475.

6 My first encounter with a volcano was with a simulacrum of Mount Etna in eruption, complete with red glow, at the Geological Museum in London. I must have been around ten years old when I first marvelled at it. N. Andrews, 'Volcanic

rhythms: Sir William Hamilton's love affair with Vesuvius', *AA Files* 60 (2010): 9–15. See also: N. Daly, 'The volcanic disaster narrative: from pleasure garden to canvas, page, and stage', *Victorian Studies* 53 (2011): 255–285.

7   He was honoured with the Royal Society's most prestigious award, the Copley Medal, in 1770. Goethe's extraordinary talents stretched to geological scholarship. He is said to have amassed nearly 18,000 rock specimens during his life. H. J. Sullivan, 'Collecting the rocks of time: Goethe, the romantics and early geology', *European Romantic Review* 10 (1999): 341–370.

8   There's little to suggest he was considered a wit, but given his disdain for rationalists, perhaps the double-entendre of 'closet' satisfied him. In her novel, Susan Sontag paints Hamilton as a 'heroic depressive' who 'ferried himself past one vortex of melancholy after another by means of an astonishing spread of enthusiasms': S. Sontag, *The Volcano Lover: A Romance* (London: Random House, 1993). Hamilton regarded 'objects and people alike [as] collectible'. His second wife, Emma, a socialite and artist's model, appears to have been no exception: G. Morson, 'Hamilton, Sir William (1731–1803), diplomatist and art collector', Oxford Dictionary of National Biography, accessed 23 June 2022, www.oxforddnb.com.

9   For example, the depiction of Etna in Athanasius Kircher's *Mundus Subterraneus* (1665). See: D. Hollis, 'Aesthetic experience, investigation, and classic ground: Responses to Etna from the First Century CE to 1773', *Journal of the Warburg and Courtauld Institutes* 83 (2020): 299–325. For more on the role of Vesuvius in the development of volcanology, see: S. Cocco, *Watching Vesuvius* (Chicago: University of Chicago Press, 2012).

10 M. A. Cheetham, 'The taste for phenomena: Mount Vesuvius and transformations in late 18th-century European landscape

depiction', *Wallraf-Richartz-Jahrbuch* 45 (1984): 131–144. Also: J. Von der Thüsen, 'Painting and the rise of volcanology: Sir William Hamilton's *Campi Phlegræi*', *Endeavour* 23 (1999): 106–109. J. Munro, 'Volcanoes', in '*True to Nature: Open-air painting in Europe 1780–1870*' (Washington: National Gallery of Art, 2020).

11 Spallanzani was a protégé (and relative) of the celebrated experimental physicist Laura Bassi (one of the first women to hold a professorship). He dressed frogs in trousers to identify the fundamentals of animal reproduction, and blinded bats to show that a sense other than vision must enable night flight. He is also known for arriving at the first correct understanding of the physics of skipping stones. There is some review of his geological contributions in: E. Vaccari, 'Lazzaro Spallanzani (1729–1799): un naturaliste italien du dixhuitième siècle et sa contribution aux sciences de la terre', *Travaux du Comité français d'Histoire de la Géologie* 3 (1996): 77–95.

12 J. Senebier, 'Historical memoir of the life and writings of Spallanzani', *The Edinburgh Magazine* (1800): 323–328, 409–416. Taken from the preface of Senebier's French translation of Spallanzani's work: L. Spallanzani, *Mémoires sur la respiration* (Geneva: Paschoud, 1803).

Jean Senebier also wrote an interesting treatise on experimentation and observation. There's a passage in it that reminded me of Haroun Tazieff's call for volcanologists to stay 'as cool as a cucumber' under pressure (p.78). Referring to two contemporaries who had 'sought truth at pain of death' on Vesuvius, Senebier explained they deserved his admiration, 'for keeping their cool in the midst of the whirlwinds of flame and smoke spewing from the volcano, and for having taken the best precautions to preserve their days ...': J. Senebier, *L'art d'observer*, (Geneva: Philibert et Chirol, 1775).

13 Homer, *The Odyssey with an English Translation by A.T. Murray, in two volumes* (London: William Heinemann, 1919).

14 R. King and S. Young, 'The Aeolian Islands: Birth and death of a human landscape', *Erdkunde* 33 (1979): 193–204.

15 J. D. Friedman *et al.*, 'Infrared surveys in Iceland in 1966', *US Geological Survey Professional Paper* 650-C (1969): C89–105.

16 They feature in a surreal conversation in Nanni Moretti's 1993 movie, *Caro Diario*.

17 Tazieff joined Cousteau aboard the *Calypso* to explore the Red Sea in 1954: H. Tazieff, *L'eau et le feu* (Paris: Arthaud, 1954).

18 C. M. M. Oppenheimer and D. A. Rothery, 'Infrared monitoring of volcanoes by satellite', *Journal of the Geological Society* 148 (1991): 563–569.

19 Another Italian highbrow, Francesco D'Arezzo, had argued as much after witnessing Mount Etna's great eruption of 1669. D'Arezzo noticed that when he pushed a stick into Etna's lava, it resisted like molten glass. He then burned nearly fifteen pounds of sulphur in Palermo, causing a public outcry as the plume of tearing gas spread. Since no similar gas arose as the lava flows inundated the outskirts of the city of Catania, how could they consist of sulphur, he reasoned? See: W. E. K. Middleton, 'The 1669 eruption of Mount Etna: Francesco d'Arezzo on the vitreous nature of lava', *Archives of Natural History* 11 (1982): 99–102.

20 See, for instance, figure 11 in: N. Métrich *et al.*, 'Paroxysms at Stromboli volcano (Italy): source, genesis and dynamics', *Frontiers in Earth Science* 9 (2021): 45. This article also discusses the rare larger explosive eruptions that rattle Stromboli.

21 In the 1862 edition of his book, Scrope even draws an analogy between explosive volcanism and uncorking a bottle of soda-water. He also likens magma to hot syrup with crystallising sugar, prefiguring the predilection of some of my colleagues for golden syrup as a lava proxy in laboratory simulations of

volcanic phenomena. (Another material I have seen used is hair gel.) G. P. Scrope, *Volcanos* (London: Longman, 1862): 42–45.

22 M. Burton *et al.*, 'Magmatic gas composition reveals the source depth of slug-driven Strombolian explosive activity, *Science* 317 (2007): 227–230.

23 Evidence for the role of crystals in Strombolian eruptions stems from experiments, field observations and studies of the ejected rock. See: J. Woitischek *et al.*, 'Strombolian eruptions and dynamics of magma degassing at Yasur Volcano (Vanuatu)', *Journal of Volcanology and Geothermal Research* 398 (2020): 106869; J. Oppenheimer *et al.*, 'Analogue experiments on the rise of large bubbles through a solids-rich suspension: a "weak plug" model for Strombolian eruptions', *Earth and Planetary Science Letters* 531 (2020): 115931; A. Caracciolo *et al.*, 'Textural and chemical features of a "soft" plug emitted during Strombolian explosions: A case study from Stromboli volcano', *Earth and Planetary Science Letters* 559 (2021): 116761.

24 J. Taddeucci *et al.*, 'High-speed imaging of strombolian explosions: The ejection velocity of pyroclasts', *Geophysical Research Letters* 39 (2012).

25 R. Cioni *et al.*, 'Understanding volcanic systems and their dynamics combining field and physical volcanology with petrology studies', in *Forecasting and Planning for Volcanic Hazards, Risks, and Disasters* (Amsterdam: Elsevier, 2021), 285–328.

26 See note 12.

27 In turn, Scrope influenced his friend Charles Lyell, whose *Principles of Geology* proved highly stimulating reading for a young Charles Darwin aboard the *Beagle*. In the 1862 edition of his book, Scrope also makes the case for 'laboratory volcanoes', suggesting Stromboli as an exemplar. True, though I can't agree with him when he says the crater can be approached and 'its

interior viewed at leisure with complete impunity'. See G. P. Scrope, *Volcanos* (London: Longman, 1862):30–31.

28   H. Davy, 'XIII. On the Phænomena of volcanoes', *The Philosophical Magazine* 4, no. 20 (1828): 85–94. Davy is mostly remembered as a chemist but he also studied and lectured on geology. See: R. Siegfried and R. H. Dott. 'Humphry Davy as geologist, 1805–29', *The British Journal for the History of Science* 9 (1976): 219–227. In a lecture he gave in London in 1812, he simulated volcanic pyrotechnics by adding water to potassium that he had placed at the top of a cone of clay and stones. See: https://www.youtube.com/watch?v=nxRHQ1xfnWc

Davy was almost overcome by gases on one fieldtrip to Vesuvius and took a month to recover. Mary Shelley drew inspiration from Davy's work while writing *Frankenstein* during the 'year without a summer', 1816, as did Jules Verne for *Journey to the Centre of the Earth*. See: A. A. Debus, 'Re-framing the science in Jules Verne's "Journey to the Center of the Earth"', *Science Fiction Studies* 33 (2006): 405–420. D. M. Pyle, 'Visions of volcanoes', *19: Interdisciplinary Studies in the Long Nineteenth Century* 25 (2017).

29   T. Monticelli and N. Covelli, *Storia de'fenomeni del Vesuvio avvenuti negli anni 1821, 1822 e parte del 1823* (Naples: Gabinetto, 1823). Monticelli, himself a resolute empiricist, generously facilitated the Vesuvius investigations of many visiting savants including Humphry Davy, Alexander von Humboldt, Charles Lyell and Christian, Crown Prince of Denmark. See: J. Brewer, 'Scientific networks, Vesuvius and politics: the case of Teodoro Monticelli in Naples, 1790–1845', *Incontri. Rivista europea di studi italiani* 34 (2019): 54–67. See also: F. Obrizzo *et al.*, 'Cultural climate in Naples between the birth and development of volcanology', *Rendiconti Online della Società Geologica Italiana* 43 (2017): 64–78.

30 Mercalli wrote a veritable textbook on volcanoes, *Vulcani Attivi della Terra*, published in 1907. The work includes dramatic photographs and first-hand accounts of the deadly eruption of Vesuvius in 1906, whose economic impacts were so great for Italy that the 1908 Olympic Games, which should have taken place in Rome, were instead hosted by London.

31 For context, see: L. Wilson and J. W. Head, 'A comparison of volcanic eruption processes on Earth, Moon, Mars, Io and Venus', *Nature* 302 (1983): 663–669.

32 For a collection of essays on Mercalli, see: M. A. Di Vito *et al.*, 'Giuseppe Mercalli da Monza al Reale Osservatorio Vesuviano: una vita tra insegnamento e ricerca', *Istituto Nazionale di Geofisica e Vulcanologia* 24 (2014). Mauro Di Vito became director of the Vesuvius Observatory in 2022.

33 Apparently, during a large eruption in 1927, lava was pouring towards the village rather than down the Sciara del Fuoco. The people put the statue of Bartholomew in front of it and the lava stopped.

What Upsets Volcanoes?

1   *Die Illustrirte Welt: Blätter aus Natur und Leben, Wissenschaft und Kunst* (Stuttgart: Hallberger, 1858): 164.

2   'Qué cosa irrita a los volcanes, que escupen fuego, frío y furia?' P. Neruda, *The Book of Questions*, translated by W. O'Daly (Port Townsend: Copper Canyon Press, 1991).

3   Of Darwin's copious *Beagle* field notes, 1,383 pages are on geological matters, whereas 368 are devoted to biology. If he referred to himself (professionally) as anything, it was as a geologist. The Sedgwick Museum in Cambridge holds his collection of igneous rocks and a number of his notebooks. See: H. E. Gruber and V. Gruber, 'The eye of reason: Darwin's development during the *Beagle* voyage', *Isis* 53 (1962): 186–200.

4 F. H. T. Rhodes, 'Darwin's search for a theory of the earth: symmetry, simplicity and speculation', *The British Journal for the History of Science* 24 (1991): 193–229. Darwin's ideas were presented to the Geological Society of London in 1838 and subsequently published in: C. Darwin, 'On the connexion of certain volcanic phenomena in South America; and on the formation of mountain chains and volcanos, as the effect of the same power by which continents are elevated', *Transactions of the Geological Society of London* 2 (1840): 601–631. Rhodes described them as 'a remarkable attempt to develop a global tectonic synthesis'.

5 L. E. Lara *et al.*, 'The AD1835 eruption at Robinson Crusoe Island discredited: Geological and historical evidence', *Progress in Physical Geography* 45 (2021): 187–206.

6 The reason why volcanoes develop at subduction zones, where old, cold oceanic plate plummets back into the mantle, is counterintuitive. The answer is water. When new ocean plate is created by undersea volcanism, it reacts with seawater circulating in the new hot crust. This hydrates minerals, turning basalt into a slippery rock called serpentinite. At a subduction zone, the oceanic plate feels increasing pressure as it sinks. Eventually, this forces out the water, which percolates into the overlying rocky layer of the Earth known as the mantle. This addition of water provokes a partial melting of the mantle, and the molten rock produced, along with the watery fluids, rises up to accumulate in the Earth's crust, where it can stew and brew for millennia before feeding an eruption at the surface.

The origins of the coastal cordillera have been much debated. See: A. Encinas *et al.*, 'Tectonosedimentary evolution of the Coastal Cordillera and Central Depression of south-Central Chile (36°30'-42° S)', *Earth-Science Reviews* 213 (2021): 103465.

7 G. Seropian *et al.*, 'A review framework of how earthquakes trigger volcanic eruptions', *Nature Communications* 12, 1004 (2021).

8 C. Mora-Stock *et al.*, 'Comparison of seismic activity for Llaima and Villarrica volcanoes prior to and after the Maule 2010 earthquake', *International Journal of Earth Sciences* 103 (2014): 2,015–2,028.

9 M. E. Pritchard *et al.*, 'Subsidence at southern Andes volcanoes induced by the 2010 Maule, Chile earthquake', *Nature Geoscience* 6 (2013): 632–636.

10 Many of their homes were ravaged by slurries of wet ash that sloughed off the mountain during rainstorms. The government proposed to relocate the population permanently, but the citizens resisted, reoccupied and rebuilt.

11 C. Wicks *et al.*, 'The role of dyking and fault control in the rapid onset of eruption at Chaitén volcano, Chile', *Nature* 478 (2011): 374–377. Cordón Caulle also erupted within two days of the Valdivia earthquake in 1960.

12 Only since the eruption has radiocarbon dating of charcoal fragments found in pumice deposits around the volcano revealed that several eruptions occurred in the past 5,000 years, and one as recently as the mid-seventeenth century: L. E. Lara *et al.*, 'Late Holocene history of Chaitén Volcano: New evidence for a 17th century eruption', *Andean Geology* 40 (2013): 249–261.

13 The volcanic clouds also spread into the Antarctic circle, where they amplified that year's spring ozone hole: S. Solomon *et al.*, 'Emergence of healing in the Antarctic ozone layer', *Science* 353 (2016): 269–274.

14 F. Arzilli *et al.*, 'The unexpected explosive sub-Plinian eruption of Calbuco volcano (22–23 April 2015; Southern Chile): triggering mechanism implications', *Journal of Volcanology and Geothermal Research* 378 (2019): 35–50.

15 F. Vilches, 'From nitrate town to internment camp: the cultural biography of Chacabuco, northern Chile', *Journal of Material Culture* 16 (2011): 241–263.

16 C. Oppenheimer, 'Mines in the sky', *Geology Today* 9 (1993): 66–68.

17 D. Blumenstiel *et al.*, 'Exposure to geogenic lithium in ancient Andeans: Unraveling lithium in mummy hair using LA-ICP-MS', *Journal of Archaeological Science* 113 (2020): 105062.

18 J. Rougier *et al.*, 'The global magnitude–frequency relationship for large explosive volcanic eruptions', *Earth and Planetary Science Letters* 482 (2018): 621–629.

19 W. J. Malfait *et al.*, 'Supervolcano eruptions driven by melt buoyancy in large silicic magma chambers', *Nature Geoscience* 7 (2014): 122–125; L. Caricchi *et al.*, 'Frequency and magnitude of volcanic eruptions controlled by magma injection and buoyancy', *Nature Geoscience* 7 (2014): 126–130.

20 M. Gardeweg and C. F. Ramírez, 'La Pacana caldera and the Atana ignimbrite—a major ash-flow and resurgent caldera complex in the Andes of northern Chile', *Bulletin of Volcanology* 49 (1987): 547–566.

21 A. S. Wilson *et al.*, 'Archaeological, radiological, and biological evidence offer insight into Inca child sacrifice', *Proceedings of the National Academy of Sciences* 110 (2013): 13,322–13,327.

22 A. S. Wilson *et al.*, 'Stable isotope and DNA evidence for ritual sequences in Inca child sacrifice', *Proceedings of the National Academy of Sciences* 104 (2007): 16,456–16,461.

23 Like the present-day residents of Talabre, the ancestors had cultivated maize and quinoa in the *quebradas* here.

24 The origin of my PhD project was a clever study by my advisors at the Open University, Peter Francis and David Rothery. Peter had worked in the central Andes since the late 1960s, and loved to be in the field. But he also recognised that for such a vast, harsh and difficult-to-access region, satellite images could provide a wealth of information on past eruptions and even ongoing activity. For Peter, the Landsat scenes were instant geological maps – there is so little vegetation in the central

Andes that it was enough to tweak the contrast of the images to differentiate rock types. What remained was to classify them. Peter pioneered this work, integrating field, laboratory and remote-sensing studies to reveal the deep histories of volcanoes.

See: P. W. Francis and D. A. Rothery, 'Using the Landsat Thematic Mapper to detect and monitor active volcanoes: An example from Láscar volcano, northern Chile', *Geology* 15 (1987): 614–617.

25 I recorded a maximum temperature of 940°C, but more importantly could see that the luminous vents would represent only tiny fractions of image pixels, as expected, verifying an approach to estimating temperatures and sizes of hot features from Landsat images. C. Oppenheimer *et al.*, 'Infrared image analysis of volcanic thermal features: Láscar Volcano, Chile, 1984–1992', *Journal of Geophysical Research: Solid Earth* 98 (1993): 4,269–4,286.

26 Rainfall has long been considered as a potential trigger of eruptions. See, for example: H. J. Johnston-Lavis, 'The relationship of the activity of Vesuvius to certain meteorological and astronomical phenomena', *Proceedings of the Royal Society of London* 40 (1886): 248–249.

27 Cases include the eruptions of Mount Ontake, Japan, in 2014, and Whakaari (White Island), New Zealand, in 2019, which claimed dozens of lives. A burst of earthquakes within a day of the 2019 disaster at Whakaari might have been a sign of the impending explosion: D. E. Dempsey *et al.*, 'Automatic precursor recognition and real-time forecasting of sudden explosive volcanic eruptions at Whakaari, New Zealand', *Nature communications* 11 (2020): 1–8.

28 S. Layana *et al.*, 'Volcanic Anomalies monitoring System (VOLCANOMS), a low-cost volcanic monitoring system based on Landsat images', *Remote Sensing* 12 (2020): 1589.

29 It's an alluring concept – the use of a common tool to monitor volcanoes around the world. But it is a far cry from an observatory embedded within the community it seeks to protect. It would also raise some thorny issues that would need much thought. For instance, if it detected unrest of a volcano, how and with whom should that information be shared? You can't go round predicting eruptions in other people's countries. M. S. Ramsey *et al.*, 'Volcanology 2030: will an orbital volcano observatory finally become a reality?', *Bulletin of Volcanology* 84 (2022): 1–8.

Emerald Isle

1 F. Perret, 'The Day's Work of a Volcanologist', *The World's Work* (November, 1907): 9,544–9,554.

2 Y. Weekes, *Volcano: A Memoir* (Leeds: Peepal Tree, 2006).

3 E. W. Freeman, 'The awful doom of St. Pierre: an extract of an account by Captain Freeman', *The Argus*, Melbourne, 11 October 1902, https://trove.nla.gov.au/newspaper/article/9088896#.

4 'Clouds of sand at sea', *New York Times*, 19 May 1902.

5 As an example of how eruption observations inform interpretation of volcanic rocks, features of some of the strata exposed on Mount Snowdon in Wales were soon recognised as products of ancient peléan eruptions: J. R. Dakyns and E. Greenly, 'III.—On the probable pelean origin of the felsitic slates of Snowdon, and their metamorphism', *Geological Magazine* 2 (1905): 541–549.

6 For a terrific review of the infamous Vesuvius eruption, see: D. M. Doronzo *et al.*, 'The 79 CE eruption of Vesuvius: A lesson from the past and the need of a multidisciplinary approach for developments in volcanology', *Earth-Science Reviews* 231 (2022): 104072.

7 Aged nineteen in 1886, Perret had dropped out of college to join Thomas Edison's laboratories. He helped to develop battery technologies, but soon left to start his own electric motors enterprise. By the late 1890s, he was constructing electric cars – sadly, a century too soon for market. As a child, he had been captivated by a coloured lithograph given to his father that depicted Pompeii's destruction. And in his teens, he witnessed the spectacular sunsets resulting from the stratospheric dust from the eruption of Krakatau in 1883. See: H. E. Belkin and T. Gidwitz, 'The contributions and influence of two Americans, Henry S. Washington and Frank A. Perret, to the study of Italian volcanism with emphasis on volcanoes in the Naples area', in *Vesuvius, Campi Flegrei, and Campanian Volcanism* (Amsterdam: Elsevier, 2020): 9–32.

My brief account here unforgivably overlooks pioneering work of others, not least French volcanologist Alfred Lacroix. See: A. Lacroix, *La Montagne Pelée et ses Eruptions* (Paris: l'Académie des Sciences sous les Auspices des Ministères de l'Instruction Publique et des Colonies, 1904).

8 Raffaele Matteucci, Director of the Royal Vesuvius Observatory, 1903–1909: '*Io amo la mia montagna, io e lei viviamo in una solitudine misteriosa e terribile, non potrei lasciarla, le sono legato per sempre; i miei pochi amici dicono che il suo respiro brucerà e farà appassire la mia povera vita*'. Quoted in: L. Civetta *et al.*, *Vesuvio negli occhi: storie di osservatori* (Unità funzionale Vulcanologia e Petrologia, 2004).

9 F. A. Perret, *The Vesuvius Eruption of 1906: Study of a volcanic cycle*, Publication 339 (Washington: Carnegie Institution, 1924): 15.

10 He considered that microphones in contact with the ground and air would provide a windfall for volcano science, and improvised field-ready equipment by cannibalising telephone receiver parts and hearing aids, connecting them to metal horns and cones.

11 Giuseppe Mercalli, who would succeed Matteucci as observatory director in 1909, also played a key role in advising the authorities and public on the prognosis of the eruption, though he and Matteucci had fallen out years earlier. Among the many geologists who visited Naples soon after the eruption was Thomas Jaggar, then assistant professor at Harvard and head of geology at MIT. He spent three weeks with Perret. Jaggar was no neophyte in volcanological matters; he had joined a scientific team on Martinique after the Pelée tragedy. With Mercalli's violent death in 1914, it was left to Perret to write the definitive monograph on the 1906 Vesuvius eruption. It is a narrative and technical masterpiece.

12 F. A. Perret, '*The Vesuvius eruption of 1906: study of a volcanic cycle*', No. 339, Carnegie Institution of Washington (1924): 46.

13 Jaggar continued: '[Perret's] books set a standard for all time for what the field science of volcanoes shall be'. T. A. Jaggar, *My Experiments with Volcanoes* (Honolulu: Hawaiian Volcano Research Association, 1956).

14 *New York Times*, 1 July 1912.

15 F. A. Perret, *The Eruption of Mt. Pelée 1929–1932*, Publication 458, (Washington: Carnegie Institution, 1935).

16 The idea that plants might register what magma is up to is not so fanciful. Diffuse carbon dioxide emissions from volcanoes can kill vegetation, and since CO_2 is the first gas to fizz out of a rising magma body, an otherwise unaccounted-for die-back of trees in a volcanic region could signal future unrest.

17 Perret would write up his account in his third monograph: F. A. Perret, *The volcano-seismic crisis at Montserrat 1933–37*, Publication 512, (Washington: Carnegie Institution, 1939). Throughout his volcanological career, he worked mostly alone without a regular salary, and never settled down in a conventional sense. He was wedded to the volcano and devoted his last years to establishing a museum on Martinique. His final major

work, published posthumously, is a grand synthesis drawing on four decades of unparalleled scrutiny of volcanic behaviour: F. A. Perret, *Volcanological Observations*, Publication 549 (Washington: Carnegie Institution, 1950).

18   See, for instance: V. C. Pinel *et al.*, 'On the relationship between cycles of eruptive activity and growth of a volcanic edifice', *Journal of Volcanology and Geothermal Research* 194 (2010): 150–164; B. G. Mason *et al.*, 'Seasonality of volcanic eruptions', *Journal of Geophysical Research* 109, B04206 (2004).

19   J. B. Shepherd *et al.*, 'Precursory activity to the 1995 eruption of the Soufrière Hills volcano, Montserrat', Seismic Research Unit, The University of the West Indies, St. Augustine, Trinidad (2002).

20   The so-called 'Wadge and Isaacs report' of 1987. A condensed version was also published: G. Wadge and M. C. Isaacs, 'Mapping the volcanic hazards from Soufriere Hills Volcano, Montserrat, West Indies, using an image processor', *Journal of the Geological Society* 145 (1988): 541–551.

21   In his tome on Pelée, Perret wrote 'next to the actual capture of the gases, spectroscopic observations seem to be the most promising method of research in this field'. Almost a century later, that's what we do.

22   Sulphur dioxide is one of the most important volcanic gases to track. For decades, this was done with the Cospec. It is such a stalwart of volcano monitoring, it even features in the 1997 disaster movie *Dante's Peak* starring Pierce Brosnan. One of the first times it was pointed at a volcano was at Masaya in the early 1970s, and it played a key role in numerous volcanic crises, including the build-up to the largest eruption of the past century, that of Pinatubo, in the Philippines, in 1991. Increasing sulphur dioxide emissions from the then almost unheard-of volcano indicated rising magma, prompting evacuations that saved tens of thousands of lives.

The Cospec was emblematic of 1960s engineering: it was practical and sturdy but also pricey and power hungry. Advances in digital imaging paved the way for a new generation of cheaper miniature ultraviolet spectrometers, and in 2001 I tried one out at Masaya with my colleagues. It worked better than I could have dreamed, and our approach is now used for routine hazard assessment at volcanoes from Vanuatu to Iceland, from Hawai'i to the Democratic Republic of Congo. It has also provided estimates for the large-scale outputs of sulphur dioxide from volcanoes, helping to understand their role in shaping the Earth's atmosphere.

See: A. S. Daag *et al.*, 'Monitoring sulphur dioxide emission at Mount Pinatubo', *Fire and Mud: Eruptions and lahars of Mount Pinatubo, Philippines* (Quezon City: Philippine Institute of Volcanology and Seismology, 1996): 409–414; D. Thompson, *Volcano Cowboys: The rocky evolution of a dangerous science* (London: Macmillan, 2002): 231; C. Oppenheimer, 'Ultraviolet sensing of volcanic sulphur emissions', *Elements* 6 (2010): 87–92.

23 *Soufrière* comes from the French *soufre* for sulphur, with a nod to the Italian Solfatara – a site of sulphurous emissions and encrustations in the Campi Flegrei near Naples. Owing to the French influence in the West Indies, several volcanoes, fumarolic areas and even villages adopted the moniker, including Soufrière St Vincent.

24 *The Times*, 17 August 1976.

25 In mid-August 1976 when the evacuation decision was made, Tazieff was uncontactable as he was on another mission – to 5,300-metre-high Sangay in Ecuador. He was on the volcano when members of a British army-led expedition were blasted with rock from a violent explosion very close to the summit. Peter Francis, my PhD advisor, was to have joined this mission at a later date. Back on Guadeloupe at the end of August, Tazieff

came close to losing his own life at the crater's edge on La Soufrière with a party of a dozen scientists. Their helmets spared them serious injuries. As the population fled their homes, another man and his cinematographers went in the opposite direction: Werner Herzog, to make his film *La Soufrière*.

26	Tazieff said to one journalist that Allègre and his team had only ever seen volcanoes in books. See this excellent summary by Francois Beauducel, former director of the Gaudeloupe volcano observatory: F. Beauducel, 'À propos de la polémique: Soufrière 1976', IPGP (http://www.ipgp.fr/~beaudu/soufriere/forum76. html). There is also a 2015 film directed by Eric Beauducel on the controversy: E. Beauducel (dir.), *Tazieff/Allègre: la guerre des volcans* (Duels, France 5: 2015).

27	H. Tazieff, 'La Soufrière, volcanology and forecasting', *Nature* 269 (1977): 96–97.

28	H. Tazieff, 'Volcanological forecasting and medical diagnosis: similarities', in *Forecasting Volcanic Events*, eds. H. Tazieff and J. C. Sabroux (Amsterdam: Elsevier, 1983), 3–7.

29	See: R. S. Fiske, 'Volcanologists, journalists, and the concerned local public: a tale of two crises in the eastern Caribbean', in *Explosive Volcanism: Interception, Evolution, and Hazard* (Washington DC: National Academy Press, 1984): 170–176.

30	Some of the tensions stemmed from poor communication. For instance, fieldworkers felt they were not always advised when the observatory seismometers registered increasing activity on the lava dome; on the other hand, staff at the observatory were unclear as to whom was authorised to enter dangerous areas. At least in this case, a solution was found – a whiteboard. Semantics were also hotly contested – one argument hinged on whether to describe the week's developments on the volcano as an 'escalation' or a 'change' in activity. Such points were not, however, pedantic, because much depended on how the observatory's pronouncements were interpreted by the community.

31 C. Oppenheimer *et al.*, 'Volcanic gas measurements by helicop-ter-borne Fourier transform spectroscopy', *International Journal of Remote Sensing* 19 (1998): 373–379.

32 In her account spanning the early years of the volcanic emer-gency, Yvonne Weekes (ibid) writes 'I am sick of hearing "and now for your daily volcano report" ... I am sick of hearing the Government telling us that we are safe ... I am sick of the noise of the [observatory] helicopter waking me up every morning.'

33 For an account of the first years of the crisis from a scientific perspective, see: R. E. A. Robertson *et al.*, 'The 1995–1998 eruption of the Soufrière Hills volcano, Montserrat, WI', *Philosophical Transactions of the Royal Society of London: Series A* 358 (2000): 1,619–1,637. For a broader view, see: P. Pattullo, *Fire from the Mountain: The Tragedy of Montserrat and the Betrayal of its People* (London: Constable, 2000). And for further perspec-tives on evacuations during volcanic crises, see: J. Barclay *et al.*, 'Livelihoods, wellbeing and the risk to life during volcanic erup-tions', *Frontiers in Earth Science* 7 (2019): 205.

34 The episode can be found online: 'Volcano! When it all Began – Montserrat Blows its Top – 1995', *Journeys with Lauren Millar*, YouTube, uploaded 4 February 2022 (https://www.youtube.com/watch?v=ESYjE2FVGT4).

35 Superimposed on this was a more gradual trend of inflation that would abruptly reset with larger collapses of the dome. The more rapid oscillations probably reflect a stop-start ascent of magma in the feeder pipe. When magma squeezed into the dome, its expansion would trigger rockfalls and small *nuées ardentes*. Gas would escape and the depressurisation would cause the summit to sag. But with gas loss, minute crystals form. This made the magma less fluid, and it stuck in the conduit, building up gas pressure and, now, causing inflation. At some threshold, the blockage cleared, resetting the cycle. See:

B. Voight *et al.*, 'Magma flow instability and cyclic activity at Soufriere Hills volcano, Montserrat, British West Indies', *Science* 283 (1999): 1,138–1,142.

36 Yvonne Weekes (ibid) writes of the 'fateful dark day' of 25 June 1997: 'At night I am haunted by the faces of the people I know who have lost their lives, their homes, their loved ones.' I still remember fondly the incorrigible Beryl Grant with her basket of fruit.

37 S. C. Loughlin *et al.*, 'Eyewitness accounts of the 25 June 1997 pyroclastic flows and surges at Soufrière Hills Volcano, Montserrat, and implications for disaster mitigation', *Geological Society, London, Memoirs* 21 (2002): 211–230.

38 Reflecting on the Soufrière Hills disaster, a UK parliamentary commission concluded in 1997 that 'Montserrat would have been immeasurably more prepared for the crisis [if the report] had been carefully read and digested ... This sorry account must be a signal example of failure in communication, political leadership and responsibility.' International Development Committee, 'International Development: First Report', Session 1997–98, 18 November 1997 (https://publications.parliament.uk/pa/cm199798/cmselect/cmintdev/267i/ido102.htm).

39 See: P. Francis, and C. Oppenheimer, *Volcanoes*, 2nd edition (Oxford: Oxford University Press, 2004).

40 One analysis has suggested the evidence available at the time was insufficient to call it either way. See: T. K. Hincks *et al.*, 'Retrospective analysis of uncertain eruption precursors at La Soufrière volcano, Guadeloupe, 1975–77: volcanic hazard assessment using a Bayesian Belief Network approach', *Journal of Applied Volcanology* 3 (2014): 1–26.

41 G. Wadge and W. P. Aspinall, 'A review of volcanic hazard and risk-assessment praxis at the Soufrière Hills Volcano, Montserrat from 1997 to 2011', *Geological Society, London, Memoirs* 39, (2014): 439–456; W. Aspinall, 'Reminiscences of a Classical

Model expert elicitation facilitator', in *Expert Judgement in Risk and Decision Analysis* (New York: Springer, 2021): 389–399.

42 I heard that on one occasion when the experts met on Montserrat for their routine risk analysis, they found the volcano subdued and were set to conclude the eruption was over when a loud explosion rattled the windows. Looking out, they saw dark ash shooting from the volcano's summit, quietly returned to their seats and revised the numbers.

43 T. A. Jaggar, 'Work of F. A. Perret on Montserrat', *The Volcano Letter*, no. 449, Department of the Interior, National Park Service (July 1937).

44 F. A. Perret, 'The Living Earth: a presentation of volcanology as a new and greater science', *Caribbean Center for Volcano Study*, Publication No. 1, 1940.

45 '. . . the non-linear complexity of volcano dynamics precludes the possibility of predicting deterministically the outcome of a volcano crisis, irrespective of the sophistication, resolution and extent of volcano monitoring instrumentation'. W. Aspinall and G. Woo, 'Counterfactual analysis of runaway volcanic explosions', *Frontiers in Earth Science* 7 (2019): 222. Put another way, anyone who claims to know with certainty what a restless or erupting volcano will do tomorrow is either deluded or a liar.

Night Market of the Ghosts

1 In *The Graphic* (21 September 1895), 360. From a sketch by W. B. D'Almeida.

2 F. W. Junghuhn, *Java, seine Gestalt, Pflanzendecke und innere Bauart*, vol. 2 (Leipzig: Arnold, 1857): 749. My own translation.

3 On one visit to the country's national volcano-monitoring centre in Bandung, I was amazed to find that seven volcanoes, scattered across the archipelago, were at the highest alert level, indicating ongoing or imminent eruption. Even a single volcanic

crisis, for instance that on La Palma in 2021, stretches human and logistical resources. Effective management of multiple simultaneous crises – some necessitating disruptive evacuations of threatened communities, others requiring additional monitoring assets – demands extraordinary preparedness, command and control.

4　P. Boomgaard, 'The high sanctuary: Local perceptions of mountains in Indonesia 1750–2000', in *Framing Indonesian Realities: Essays in Symbolic Anthropology in Honour of Reimar Schefold* (Leiden: KITLV Press, 2003): 295–314. Boomgaard writes: 'The "holy trinity" of mountain, forest, and water, with the mountain as the supreme deity, symbolized, and in some areas still symbolizes, the forces that make life possible.'

5　Nor was I aware then that the traditional landowners, the *Puku*, hold a ceremony to honour the ancestors and earth spirits at a cave on the cone's flank, laying offerings of cotton wound with red thread (symbolising smoke and lava), tobacco and rice, before slaughtering a goat with bare hands. The performance ends with a ritual meal and the thrusting of an iron lance into the ground. These acts bind and protect the community. See U. U. Frömming, 'Volcanoes: symbolic places of resistance. Political appropriation of nature in Flores, Indonesia', *Violence in Indonesia* (Hamburg: Abera Verlag, 2001): 270–281.

6　Demon is, in fact, a clan name.

7　For a brilliant biography of Junghuhn, see: R. Sternagel, *Der Humboldt von Java: Leben und Werk des Naturforschers Franz Wilhelm Junghuhn 1809–1864*, (Halle: Mitteldeutscher Verlag, 2013).

8　The article leads with an epigraph taken from Goethe's *Faust*, foreshadowing the amalgam of transcendentalism and empiricism that characterises much of Junghuhn's oeuvre: 'Inscrutable at broad day, nature does not suffer her veil to be torn from her; and what she does not choose to reveal to thy mind, thou wilt not wrest from her by levers and screws.' J.W. Von Goethe, trans. A. Hayward, *Faust:A Dramatic Poem* (London: E. Moxon, 1834.)

9 I was part of a team that dated the Laacher See eruption with astonishing precision through radiocarbon measurements of trees carbonised by the *nuées ardentes*: F. Reinig *et al.*, 'Precise date for the Laacher See eruption synchronizes the Younger Dryas', *Nature* 595 (2021): 66–69.

10 Sternagel, ibid, 67.

11 U. Bosma, 'Franz Junghuhn's three-dimensional and transcendental Java', in *The Role of Religions in the European Perception of Insular and Mainland Southeast Asia* (Cambridge: Cambridge Scholars Publishing, 2016): 175–206.

12 Architecturally, both Prambenan and Borobudur temples were designed as reflections of sacred mountains and the cosmic order.

13 Sternagel, ibid, 50.

14 E. M. Beekman, 'Junghuhn's perception of Javanese nature', *Canadian Journal of Nederlandic Studies* 12 (1991): 136.

15 Junghuhn was also an ardent critic of colonialism, and a defender of universal human freedoms. He wrote an astonishing dialectic on colonialism and religion, which was so controversial that his authorship was only disclosed after his death: F. W. Junghuhn, *Licht-en Schaduwbeelden uit de Binnenlanden van Java* (Amsterdam: F. Günst, 1867).

16 F. W. Junghuhn, *Java-Album: Landschafts-Ansichten von Java nach der Natur aufgenommen* (Leipzig: Arnoldische Buchhandlung, 1856). Some volcanologists still cite his research today. For instance, the accuracy and detail of a sketch of the crater of Gede, which Junghuhn witnessed in eruption in 1843, provided historical perspective for: A. Belousov *et al.*, 'Volcaniclastic stratigraphy of Gede Volcano, West Java, Indonesia: how it erupted and when', *Journal of Volcanology and*

Geothermal Research 301 (2015): 238–252. Contemporaries such as Scrope were also cognisant and complimentary of Junghuhn's observations, but his reputation was eclipsed, in Britain, certainly, by Alfred Russel Wallace, and very little of his work has ever been translated into English.

17 Junghuhn's most iconic illustration is based on sketches from his first expeditions, in 1836, to Merapi. These may be the earliest accurate depictions of a lava dome: F. Junghuhn, *Topographische und naturwissenschaftliche Reise durch Java* (Madgeburg: Emil Baensch, 1845).

18 E. D. Inandiak and H. Dono, *Merapi Omahku* (Babad Alas: Yogyakarta 2010).

19 The importance of this cannot be downplayed – science and technology are insufficient by themselves to prevent disasters. The disaster following a landslide of Krakatau into the sea in 2018 illustrates the point. The results of a study published six years earlier had demonstrated the threat of just such a partial collapse of the cone. Based on a computer simulation, it predicted tsunami wave heights of a few metres along the Sunda coastline in the event of such a scenario. The authors concluded: 'Owing to the high population … the tsunami might present a significant risk. However, as the travel time of the tsunami is several tens of minutes … a rapid detection of the collapse by the volcano observatory, coupled with an efficient alert system on the coast, could prevent this hypothetical event from being deadly'. The science did not translate into effective action, however, and more than 400 people were killed on 22 December 2018. The correspondence between modelled and actual tsunami characteristics is remarkably close. T. Giachetti *et al.*, 'Tsunami hazard related to a flank collapse of Anak Krakatau Volcano, Sunda Strait, Indonesia', *Geological Society, London, Special Publications* 361, (2012): 79–90. L. Ye *et al.*, 'The 22 December 2018 tsunami from flank

collapse of Anak Krakatau volcano during eruption', *Science Advances* 6, eaaz1377 (2020).

20 Thirty-four others perished with him, including a journalist. A larger blast during the night of 4–5 November claimed well over 100 lives. For an account of the eruption, see: Surono *et al.*, 'The 2010 explosive eruption of Java's Merapi volcano – a '100-year event', *Journal of Volcanology and Geothermal Research* 241 (2012): 121–135.

21 Reflecting on changing attitudes towards volcanoes in Indonesia, Boomgaard (ibid) noted: 'tourists rush in where once believers feared to tread'.

22 He confided that, as the activity escalated, he had resisted great pressure to meet with Mbah Maridjan: 'It would have changed nothing and just been entertainment for the newspapers,' he told me. 'I was much more concerned with deciding the evacuation zones. It was terrible pressure.' I met Surono in Kinahrejo during the *labuhan* in 2016 – I had invited him to join us during filming of *Into the Inferno*. When he was spotted, he was accorded the honour of introducing the all-night shadow puppet performance, *wayang kulit*, accompanied by a full *gamelan* orchestra.

For more on Merapi's extraordinary complexity, see: A. Bobbette, '*The pulse of the Earth: political geology in Java*', (Durham, NC: Duke University Press, 2023).

23 T. S. Raffles, *The History of Java* (London: Black, Parbury & Allen, 1817); T.S. Raffles, *Memoir of the life and public services of Sir Thomas Stamford Raffles, F.R.S. &c., particularly in the government of Java, 1811–1816, and of Bencoolen and its dependencies, 1817–1824: with details of the commerce and resources of the eastern archipelago, and selections from his correspondence* (London: John Murray, 1830). There is also the record in the rocks: H. Sigurdsson and S. Carey, 'Plinian and co-ignimbrite tephra fall from the 1815 eruption of Tambora volcano', *Bulletin of Volcanology*, 51 (1989): 243–270.

24 J. Crawfurd, *A Descriptive Dictionary of the Indian Islands and Adjacent Countries* (London: Bradbury and Evans, 1856).

25 I. Suhendro *et al.*, 'Magma chamber stratification of the 1815 Tambora caldera-forming eruption', *Bulletin of Volcanology* 83, no. 63 (2021).

26 This poses a critical question for contemporary volcanic risk management. The 5 April explosion was itself comparable to the 18 May 1980 eruption of Mount Saint Helens in the United States. Imagine experiencing an eruption of that scale, but being unaware it is just the throat-clearing event preparing the way for something a hundred times larger. In such a scenario, all ground-monitoring equipment near the volcano will be destroyed – how could volcanologists recognise that such a colossal event is a mere appetizer for the main course?

27 Especially sappanwood, once widely traded for its medicinal benefits and as a red dye. This rosy picture, though, overlooks the fact that for labourers life was surely harsh owing to the excesses of the island's native nobility and external greed for products.

28 P. R. Goethals, *Aspects of Local Government in a Sumbawan village (Eastern Indonesia)*, Cornell Modern Indonesia Project (New York: Cornell University, 1961).

29 M. J. Hitchcock, 'Is this evidence for the lost kingdoms of Tambora?', *Indonesian Circle* 33 (1984): 30–35.

30 The first concrete steps towards understanding the connections between volcanism and climate were made following the eruption of Krakatau in 1883, a disaster that claimed 36,000 lives. People around the world remarked on vivid sunsets and prolonged twilights, among them Reverend Sereno Bishop in Honolulu, Hawaii. He wrote of a white aureole around the sun and a diffusion of light by fine particles, an effect named after him: 'Bishop's rings'. In 1888, the Royal Society of London

published a landmark study of the global effects of Krakatau's haze, drawing on worldwide reports of dry fogs, red twilights and blue suns following known volcanic eruptions back to 1500. I think of it as the first product of 'crowd science' in volcanology. *The Eruption of Krakatoa, and Subsequent Phenomena: Report of the Krakatoa Committee of the Royal Society*, ed. G. J. Symons (London: Trübner, 1888).

The eruption of Katmai in Alaska in 1912 shed further light on the matter: C. G. Abbot and F. E. Fowle, 'Volcanoes and climate', *Smithsonian Miscellaneous Collections* (1913). Later that decade, the American meteorologist William Humphreys made a more systematic assessment of the impacts of global volcanism on climate: W. J. Humphreys, 'Factors of climatic control', *Journal of the Franklin Institute* 188 (1919): 775–810 and 189 (1920): 63–98.

Hubert Lamb, a British meteorologist inspired by studies that emerged after the 1963 eruption of Agung volcano in Bali, and let loose in the dusty archives of the Meteorological Office in London in the 1960s, had more data to hand, and wrote an immense paper that introduced the first measure of an eruption's climate-changing potential: the 'dust veil index': H. H. Lamb, 'Volcanic dust in the atmosphere; with a chronology and assessment of its meteorological significance', *Philosophical Transactions of the Royal Society of London: Series A, Mathematical and Physical Sciences* 266 (1970): 425–533. On Agung, see: S. C. Mossop, 'Stratospheric particles at 20 km altitude', *Geochimica et Cosmochimica Acta* 29 (1965): 201–207.

31 R. F. Pueschel, 'Stratospheric aerosols: Formation, properties, effects', *Journal of Aerosol Science* 27 (1996): 383–402.

32 Whenever I give a talk on volcanoes and climate, I always show a sequence of satellite images revealing dust in the stratosphere from Pinatubo. Nothing better reveals how volcanic aerosol can envelop the entire planet – and how long it stays up there. I

usually hear gasps of astonishment from the audience as the animation runs. I also show a photo taken by a Space Shuttle astronaut over South America a few months after the eruption. Two dusky layers stand out, more than twelve miles above the surface. It's then easy to imagine the dust filtering out some of the sunlight falling on the Earth and how that must cool the climate. Additional observations from satellites and weather observatories show the impacts on global temperatures, including a pronounced summertime cooling across swathes of North America and Eurasia in 1992. See: V. Aquila *et al.*, 'Impacts of the eruption of Mount Pinatubo on surface temperatures and precipitation forecasts with the NASA GEOS subseasonal-to-seasonal system', *Journal of Geophysical Research: Atmospheres* 126 (2021): e2021JD034830.

One of the ideas to counter global warming is to simulate the effect of volcanic aerosol, for instance by burning sulphur in the stratosphere, but there are numerous arguments not to go down this path. See: A. Robock, '20 reasons why geoengineering may be a bad idea', *Bulletin of the Atomic Scientists* 64 (2008): 14–18.

33 C. U. Hammer *et al.*, 'Greenland ice sheet evidence of post-glacial volcanism and its climatic impact', *Nature* 288 (1980): 230–235.

34 Ice cores can be dated a bit like tree rings, by identifying seasonal layers. The sulphur alone cannot pinpoint the source volcano, but sometimes it is accompanied by volcanic ash particles, which can be analysed geochemically to provide a signature for which a match with a known volcano can be sought. One of the earliest studies of this method was made by my Erebus colleague Phil Kyle: P. R. Kyle and P. A. Jezek, 'Compositions of three tephra layers from the Byrd Station ice core, Antarctica', *Journal of Volcanology and Geothermal Research* 4 (1978): 225–232. Ice core records of volcanism are fabulous now – see, for example: M. Sigl *et al.*, 'Volcanic stratospheric sulphur injections and

aerosol optical depth during the Holocene (past 11,500 years) from a bipolar ice-core array', *Earth System Science Data* 14 (2022): 3,167–3,196.

35 The relation between the quantity of sulphate aerosol in the stratosphere and climatic change is complex and non-linear. Very high amounts of sulphur are thought to result in formation of larger aerosol particles, which are less effective at filtering out sunlight and settle out of the stratosphere more rapidly. So, the surface cooling effect should approach some kind of a limit.

36 C. A. Stephens, '1800 and Froze to Death', in *The Children's Book of Thanksgiving Stories*, ed. A. D. Dickinson (New York: Doubleday, Page & Company, 1915).

37 V. Bright, 'Black Harris, Mountain Man, Teller of Tales', *Oregon Historical Quarterly* 52 (1951): 3–20. Bright writes: 'Numerous of these tales have survived to this day, and many a factual history of events is embellished with the fanciful dreamings of such men as Black Harris'. The yarn in question, attributed to Harris, is recounted in: E. Bennett, *The Prairie Flower; or Adventures in the Far West* (Cincinnati: U. P. James, 1852): 29.

38 I've drawn here on an illuminating and offbeat paper by journalist, magician and local historian June Barrows Mussey (alias Henry Hay): B. Mussey, 'Yankee chills, Ohio fever', *New England Quarterly* (1949): 435–451. Mussey, aware of the teleconnection with Tambora, wrote lyrically: 'that thin chain of cause and effect seems almost like superstition'. See also: H. M. Stommel and E. Stommel, *Volcano Weather: the story of 1816, the year without a summer* (Newport, Rhode Island: Seven Seas Press, 1983).

39 A sense of ordinary people's despair echoes in this account from North Carolina, where the grain fields also languished because of the volcano weather: 'It was with sorrowful and troubled hearts that we gathered our second crop of hay and our corn crop, which were so scanty that we reaped only a third of what we usually get, and wondered how we could subsist until next

year's harvest.' 'Records of the Moravians in North Carolina 1752–1879', ed. A. L. Fries, (Raleigh: State Department of Archives and History, vol. VII, 1947): 3,294–3,318.

40 At least the British could draw comfort from Napoleon's exile on the remote volcanic island of St Helena in the South Atlantic ocean.

41 C. von Clausewitz, *Politische Schriften und Briefe*, ed. Hans Rothfels (Munich: Drei Masken Verlag, 1922): 189–191.

42 R. Glaser *et al.*, 'Climate of migration? How climate triggered migration from southwest Germany to North America during the 19th century', *Climate of the Past* 13 (2017): 1,573–1,592.

43 Starving protestors in the towns of Littleport and Ely, not far from my hometown of Cambridge, were arrested and five hanged on 28 June 1816 to serve as an example to others who might be tempted to take to the streets.

44 J. D. Post, 'The economic crisis of 1816–1817 and its social and political consequences', *The Journal of Economic History* 30 (1970): 248–250.

45 Clearly, over the past century, many scholars from diverse fields have thought and written about Tambora and its long-range impacts. A whirl of publishing accompanied the bicentenary of the eruption. For me, though, the seminal work remains the historian John Post's classic: J. D. Post, *The Last Great Subsistence Crisis in the Western World* (Baltimore: Johns Hopkins University Press, 1977). For a dramatised version of events, see: G. Glasfurd, *The Year Without Summer* (London: John Murray Press, 2020). For more on volcano forensics and other case studies, see: C. Oppenheimer, *Eruptions That Shook the World* (Cambridge: Cambridge University Press, 2011).

46 I've thought a lot about how far we can push interpretations of history and how confident we can be in them. For every statement such as 'the Tambora eruption contributed to the rise of Fascism in Europe', there is a counterfactual along the lines of

'Fascism may not have reared its head at this time had Tambora not erupted when it did'. Turning the statement around highlights the multitude of other factors – social, political, economic, cultural – likely to be implicated, too. I feel the best we can do is attempt to offer the most rational arguments in the light of the often very diverse strands of evidence available to us. Those arguments may end up being substantiated or negated but likely never proven. I came to this viewpoint while considering explanations of contemporary world affairs. The Arab Spring strikes me as a good example. Some have linked the uprisings, which got underway in 2010, to climate change, food insecurity in China and soaring prices of staples in northern Africa. Others see failures of governance as driving factors. If we cannot know for sure the causes of events that happened under our noses, as it were, in our own time, with all the immensity of information to hand, how much more challenging it is to impute causality to historical episodes dating back centuries or millennia. J. Schilling *et al.*, 'Climate change vulnerability, water resources and social implications in North Africa', *Regional Environmental Change* 20(15) (2020).

47 Toba's ash has been traced as far as eastern and southern Africa. See: E. I. Smith, *et al.*, 'Humans thrived in South Africa through the Toba eruption about 74,000 years ago', *Nature* 555 (2018): 511–515.

48 His was the first detailed ethnographic, cartographic and botanical survey of the district: F. W. Junghuhn, *Die Battaländer auf Sumatra: Im Auftrage Sr. Excellenz des General-Gouverneurs von Niederländisch-Indien Hrn. P. Merkus in den Jahren 1840 und 1841* (Berlin: Druck und Verlag von G. Reimer, 1847).

49 A popular scenario assumes Toba emitted 100 times as much sulphur as did Pinatubo in 1991. See for example: A. Robock *et al.*, 'Did the Toba volcanic eruption of ~74 ka BP produce widespread glaciation?', *Journal of Geophysical Research: Atmospheres*

114 (2009); B. A. Black *et al.*, 'Global climate disruption and regional climate shelters after the Toba supereruption', *Proceedings of the National Academy of Sciences* 118 (2021); C. Timmreck *et al.*, 'Climate response to the Toba super-eruption: Regional changes', *Quaternary International* 258 (2012): 30–44.

While the models may give reasonable insights into the impacts of such a sulphur loading, I have long argued they have limited relevance for understanding the Youngest Toba Tuff eruption until we have a better handle on the sulphur output. Furthermore, climate impacts of volcanic eruptions depend on the prevailing climate state (sensitivity to forcing, atmospheric circulation), which we cannot fix in this case. Where Toba fall-out is found in seasonally resolved sedimentary deposits (as in Jawalapuram in India), then there is a basis for interpreting aspects of the eruption's climate and environmental impacts. Identifying Toba in polar ice core records would represent a great advance; progress is being made: L. Crick *et al.*, 'New insights into the ~74 ka Toba eruption from sulphur isotopes of polar ice cores', *Climate of the Past* 17 (2021): 2,119–2,137.

50   Estimated around 34,000 cubic kilometres. A few days later, I was getting lost in dense forest trying to find a way up Sibayak volcano, just north of Lake Toba. It was precisely 100 years after the 1883 eruption of Krakatau, which I knew of from Peter Francis' *Volcanoes* book. I came across a sulphur miner and accompanied him up a trail to the top. For the figure on the region containing melt beneath Toba today, see: J. Stankiewicz *et al.*, 'Lake Toba volcano magma chamber imaged by ambient seismic noise tomography', *Geophysical Research Letters* 37 (2010).

51   For a forceful rejection of the interpretations and arguments, see: P. Mellars *et al.*, 'Genetic and archaeological perspectives on the initial modern human colonization of southern Asia', *Proceedings of the National Academy of Sciences* 110 (2013): 10,699–10,704. Discussion of the migration of *Homo sapiens*

from Africa has to account for increasingly early dates for the presence of the species in Australia. See: C. Clarkson *et al.*, 'Human occupation of northern Australia by 65,000 years ago', *Nature* 547 (2017).

52 S. H. Ambrose, 'Late Pleistocene human population bottle-necks, volcanic winter, and differentiation of modern humans', *Journal of Human Evolution* 34 (1998): 623–651.

53 C. Lane *et al.*, 'Ash from the Toba supereruption in Lake Malawi shows no volcanic winter in East Africa at 75 ka', *Proceedings of the National Academy of Sciences* 110 (2013): 8,025–8,029; C. Clarkson, *et al.*, 'Human occupation of northern India spans the Toba super-eruption ~74,000 years ago', *Nature Communications* 11, no. 961 (2020).

54 J. Rougier *et al.*, 'The global magnitude–frequency relationship for large explosive volcanic eruptions', *Earth and Planetary Science Letters* 482 (2018): 621–629.

The last super-eruption occurred at Taupō: C. J. N. Wilson *et al.*, 'The 26·5 ka Oruanui eruption, Taupo volcano, New Zealand: development, characteristics and evacuation of a large rhyolitic magma body', *Journal of Petrology* 47, (2006): 35–69. (The date has since been revised to circa 25,600 years ago.)

55 R. L. Christiansen *et al.*, *'Preliminary assessment of volcanic and hydrothermal hazards in Yellowstone National Park and vicinity'*, U. S. Geological Survey, Open-file report 1071 (2007).

So, it is more likely that some other volcano will launch the next super-eruption. It could well be one of Indonesia's, given the country's plenitude in fire mountains. But I do not believe volcanoes will destroy the world, for two reasons. Firstly, although the societal consequences of such a high magnitude event may be very severe regionally, and amplified worldwide by vulnerabilities in the global food system, the climatic effects will not shut down agriculture altogether, and they will diminish over a period of years. Secondly, and rather sadly, I think social media, demagoguery and other existential threats are more likely to undo the human enterprise before volcanoes get another go. See: C. Sagan, 'Nuclear war and climatic catastrophe: Some policy implications', *Foreign Affairs* 62 (1983): 257–292. A. E. Snyder-Beattie *et al.*, 'An upper bound for the background rate of human extinction', *Scientific Reports* 9, no. 11054 (2019). A. Donovan and C. Oppenheimer, 'Imagining the unimaginable: communicating extreme volcanic risk', in *Observing the Volcano World* (New York: Springer, 2016): 149–163. H. Lin, 'The existential threat from cyber-enabled information warfare', *Bulletin of the Atomic Scientists* 75 (2019): 187–196.

56 Based on several lines of geochemical and isotopic evidence in the rocks. See: A. E. Mucek *et al.*, 'Post-supereruption recovery at Toba Caldera', *Nature Communications* 8, no. 15,248 (2017).

White Mountain, Heaven Lake

1   C. Kim, 'Taedong yŏjido' [1864]. Original image from Harvard-Yenching Library, Harvard College Library, Harvard University. https://iiif.lib.harvard.edu/manifests/view/drs:15388328$14i

2   The 'First Lady' of DPRK is quoted as saying this atop Mount Paektu in 2018. H. Shin and J. Lee 'Fulfilling a dream, South Korea's Moon visits sacred North Korean mountain with Kim', *Reuters*, https://www.reuters.com/article/us-northkorea-south-korea-summit-mountain-idUSKCN1M006F [2018].

3   Earlier that year, 2018, Kim Jong Un had crossed the military demarcation line that divides the countries, the first DPRK leader to do so since 1952. As a symbol of peace, he and Moon Jae-In ceremonially planted a pine tree, placing around it soil from both Mount Paektu and Mount Halla volcano (on Jeju Island in South Korea). It's astonishing to me that a mountain ripped apart in one of the most dramatic eruptions in history has become a symbol both for revolutionary defiance and

healing – a geopolitical reflection of the Jekyll-and-Hyde nature of volcanoes.

4   H. Machida *et al.*, 'Historical eruptions of the Changbai volcano resulting in large-scale forest devastation (deduced from widespread tephra)', in H. Yang *et al.*, '*The temperate forest ecosystem*', ITE symposium NERC/ITE, Cumbria (1987), 23–26.

5   The thickness of an ash deposit falls off more or less exponentially with distance away from the volcanic vent.

6   Another benchmark study was performed by Sarah Horn and Hans-Ulrich Schmincke, who did fieldwork on Paektu both sides of the border: S. Horn, and H. U. Schmincke, 'Volatile emission during the eruption of Baitoushan Volcano (China/North Korea) ca. 969 AD', *Bulletin of Volcanology* 61 (2000): 537–555. They estimated the size and volatile release of the eruption and its date from radiocarbon measurements.

7   I've drawn on numerous sources, including: K. Kwon and P. H. Kim, 'Will the dormant volcano erupt again? Mt. Paektu and contemporary Sino-Korean relations', *Asia-Pacific Journal-Japan Focus* 17 (2019); V. Ten, 'Mt. Paektu and Sŏndo (仙道 the way of immortality) in contemporary South Korea: The case of GiCheon (氣天)', *Situations* 10 (2017): 145–170; V. Ten and R. Winstanley-Chesters, *New goddesses on Mt. Paektu: Transformation, myth and gender in Korean landscape*, (2020), http://dx.doi.org/10.2139/ssrn.3755038; A. Schmid, *Korea Between Empires, 1895–1919*, (New York: Columbia University Press, 2002); K. Pratt, 'Portrait of a volcano: the paradox of Paektu (Changbaishan)', in *TephroArchaeology in the North Pacific*, eds. G. L. Barnes and T. Soda (Oxford: Archaeopress, 2019): 114–139; A. Donovan, 'Politics of the lively geos: volcanism and geomancy in Korea', in *Political Geology* (London: Palgrave Macmillan, 2019): 293–343; S. Haeussler, 'Descriptions of the Baekdusan and the surrounding area in Russian and German travel accounts', *The Review of Korean Studies* 13 (2010):

151–186; R. Rogaski, *Knowing Manchuria: environments, the senses, and natural knowledge on an Asian borderland* (Chicago: University of Chicago Press, 2022).

8 E. R. Canda, 'The Korean mountain spirit', *Korean Journal* 20-9 (1980): 11–16.

9 H. E. M. James, *The Long White Mountan: or, a journey in Manchuria, with some account of the history, people, administration and religion of that country* (London: Longmans, Green, and co., 1888): 454.

10 Z. Shen and X. Yafeng, 'Contested border: A historical investigation into the Sino-Korean border issue, 1950–1964', *Asian Perspective* 37 (2013): 1–30. Today, around a quarter of the volcano is in North Korea. Despite ceding territory, there has been a strong push in China on its side of the border to strengthen Chinese identity in what is a largely ethnically Korean community.

11 I had even (incorrectly) estimated its age: C. Oppenheimer, 'Ice core and palaeoclimatic evidence for the timing and nature of the great mid-13th century volcanic eruption', *International Journal of Climatology* 23 (2003): 417–426.

12 F. Wang et al., 'Influence of the March 11, 2011 M_w 9.0 Tohoku-oki earthquake on regional volcanic activities', *Chinese Science Bulletin* 56 (2011): 2,077–2,081.

13 J. Xu et al., 'Recent unrest of Changbaishan volcano, northeast China: A precursor of a future eruption?', *Geophysical Research Letters* 39, L16305 (2012).

14 There have also been rumours that North Korean nuclear tests might destabilise the volcano, though these seem to verge on the intentionally scurrilous rather than scientifically based. One study led by South Korean seismologists concluded: 'North Korean nuclear explosions are expected to produce pressure changes of tens to hundreds of kilopascals, causing concern over the possible triggering of volcanic eruption.' However,

analysis of the seismicity of Mount Paektu after past tests has revealed nothing. T. K. Hong *et al.*, 'Prediction of ground motion and dynamic stress change in Baekdusan (Changbaishan) volcano caused by a North Korean nuclear explosion', *Scientific reports* 6, no. 21477 (2016); G. Liu *et al.*, 'Detecting remotely triggered microseismicity around Changbaishan Volcano following nuclear explosions in North Korea and large distant earthquakes around the world', *Geophysical Research Letters* 44 (2017): 4,829–4,838.

15 R. Stone, 'Vigil at North Korea's Mount Doom', *Science* 334 (2011): 584–588; R. Stone, 'Sizing up a slumbering giant', *Science* 341 (2013): 1,060–1,061.

16 The Pyongyang International Information Centre of New Technology and Economy (PIINTEC) was established in 2003 to foster knowledge exchange in science and technology and build international partnerships with DPRK institutions. One of its main branches is the Environmental Information Media Centre, with which we work. Their remit includes raising awareness of environmental issues and sustainable development (green economy, water resources, environmental protection, ecosystem restoration).

17 One instruction I received read: 'Dr Clive to carry the necessary tools for geological research in the field including the rope ladder for climbing up and down.' I found a three-storey escape ladder through a fire-safety retailer, but later realised something ten times longer was being sought to access the vertical wall of Paektu's crater. That sounded terrifying and I quietly dropped the matter.

18 There is a remarkable clip of its unveiling shown in *Into the Inferno*, dir. W. Herzog (Netflix, 2016).

19 F. E. Younghusband, '*The heart of a continent: a narrative of travels in Manchuria, across the Gobi desert, through the Himalayas, the Pamirs, and Chitral, 1884–1894*. (London: John Murray, 1896): 15.

20 Both China and the DPRK have UN-designated biosphere reserves on the volcano that abut each other at the international border. Despite the protection afforded by the reserves, the ecosystem is changing as the climate warms and snow melts earlier in the year. S. Zong *et al.*, 'Upward range shift of a dominant alpine shrub related to 50 years of snow cover change', *Remote Sensing of Environment* 268 (2022): 112773.

21 See: C. M. Crisafulli and V. H. Dale, *Ecological Responses at Mount St. Helens: Revisited 35 years after the 1980 Eruption* (New York: Springer, 2018).

22 I. Thornton, *Krakatau: the destruction and reassembly of an island ecosystem* (Harvard: Harvard University Press, 1997).

23 Kim Ju Song was especially keen to investigate the eruption history. 'This will help us predict the next eruption,' he explained. 'We expect it will be smaller than the Millennium Eruption, but we have to plan for the worst case. The rhythm of the volcano suggests there could be a small eruption in our generation.' As we had heard on our previous visit, there was still a dearth of reliable modern monitoring equipment and data processing software: 'We cannot import it because of sanctions. And because of the intensification of the situation in Korea, people are afraid to invest in science. A Japanese geologist promised core research but was not allowed by his government to come to DPRK and the equipment he promised was blocked. As a scientist he was eager to collaborate but sanctions prevented this.'

24 For a picture of the vegetation, see: J. W. Kim *et al.*, 'Alpine vegetation on the Paekdu-san (Changbaishan) summit of the north-east China', *Journal of Plant Biology* 62 (2019): 436–450.

25 There have been a few accidental encounters with magma during geothermal drilling, notably one in 2009 in Iceland at Krafla volcano. The magma quickly solidified in the drill hole forming a plug – there was no runaway eruption! But the idea of drilling into magma to depressurize a whole reservoir of the

stuff is very far-fetched in my opinion (but perhaps I just can't think outside the box on this one). There is an ambitious proposal for scientific drilling into the Krafla chamber, the Krafla Magma Testbed – 'a journey to molten Earth' as they describe it on their website: https://www.kmt.is/

See also: J. Eichelberger, 'Planning an international magma observatory', *Eos, 100*, doi:10.1029/2019EO125255 (2019).

26   Tensions were running high at the time, following the wounding of two South Korean soldiers by landmines south of the DMZ. In retaliation, Seoul had started belting out propaganda from massive loudspeakers aimed across the border. There had just been an exchange of shells, thankfully without casualties.

27   K. S. Ri *et al.*, 'Evidence for partial melt in the crust beneath Mt. Paektu (Changbaishan), Democratic People's Republic of Korea and China', *Science advances* 2 (2016): e1501513. James strengthened the observations in a follow-up study in collaboration with Chinese and DPRK colleagues: J. O. S. Hammond *et al.*, 'Distribution of partial melt beneath Changbaishan/Paektu Volcano, China/democratic people's Republic of Korea', *Geochemistry, Geophysics, Geosystems* 21 (2020): e2019GC008461.

28   F. Miyake *et al.*, 'A signature of cosmic-ray increase in AD 774–775 from tree rings in Japan', *Nature* 486 (2012): 240–242.

29   Given such severe solar activity can intensify aurora and extend them to lower latitudes, it may be no coincidence that the prominent Japanese Buddhist priest Kūkai, born in 774 CE, was named Henjō Kōngō, which translates as 'worldwide shining diamond'. See: U. Büntgen, *et al.*, 'Tree rings reveal globally coherent signature of cosmogenic radiocarbon events in 774 and 993 CE', *Nature Communications* 9, no. 3605 (2018).

30   We also studied the rings of trees from around the Northern Hemisphere to see if there was any indication of summer cooling. Despite the eruption's size, there is little evidence it had any significant impacts on climate, probably due to a low sulphur

output and winter timing. See: C. Oppenheimer *et al.*, 'Multi-proxy dating the 'Millennium Eruption' of Changbaishan to late 946 CE', *Quaternary Science Reviews* 158 (2017): 164–171.

Lava Floods and Hurtling Flames

1   C. Strahlheim (J. K. Friederich), '*Die Wundermappe oder sammtliche Kunst- und Natur-Wunder des ganzen Erdballs*' (Frankfurt: Comptoir für Literatur und Kunst, 1837).

2   Werner Herzog, personal communication, 2015.

3   Despite being utterly devoid of sunlight so far below sea level, the ridge still teems with life: at the base of the food chain are bacteria nourished by hot, mineralised fluids venting from fissures; at the top, a frenzy of weird molluscs and crustaceans.

4   The Earth is hot inside from primordial bombardments by asteroids and radioactivity. The only large region of liquid in the interior is the iron-rich outer core. Above, the silicate mantle is solid, but it can still deform and convect, with hotter regions rising and colder ones sinking. Most of Earth's volcanoes lie on the seabed, and they exist because the ascending mantle starts melting at a depth of around 100 km due to the reducing weight (and pressure) of overlying rock. But only the minerals with the lowest melting points liquify, generating droplets of basalt. The molten rock is focused by pressure upwards and accumulates a few kilometres below the seabed to form magma chambers, which feed the eruptions at mid-ocean ridges.

   Mantle plumes similarly feed volcanoes through this 'decompression melting', but more prodigiously because their higher temperatures produce more molten rock. Volcanoes appear where mantle plumes blowtorch through tectonic plates – this is how the Hawaiian Islands and the trail of seamounts that stretch to the north formed over the last seventy million years. When mantle plumes impinge on continents, they can cause rifting, as seen in eastern Africa. If sustained, the stretching can eventually break apart the continent and heal the rift with new ocean floor, as it did in the Red Sea and North Atlantic. Iceland is considered to be the end result of hotspot volcanism that originally began around sixty million years ago, shortly before the opening of the North Atlantic.

See: K. Yuan and B. Romanowicz, 'Seismic evidence for partial melting at the root of major hot spot plumes', *Science* 357 (2017): 393–397.

5 It was the night of 24 June 1982, and flight BA009 from Kuala Lumpur to Perth was at 37,000 feet. There were 247 passengers on board and nearly 100 tonnes of fuel. The full story of Captain Eric Moody's calm in a terrifying crisis makes for excellent reading: See: J. Diamond, 'Down to a sunless sea: the anatomy of an incident', *The Log*, April 1986: 13–16.

6 The Volcanic Ash Advisory Centres (VAACS) are located in Anchorage, Buenos Aires, Darwin, London, Montreal, Tokyo, Toulouse, Washington DC and Wellington. Their situation within meteorological agencies is a reminder of the entwined histories of meteorology, volcanology and climatology.

7 Despite this infamy, Icelandair, the national carrier, proudly names its aircraft after the country's volcanoes. I've flown on *Öræfajökull* and *Hekla*, and there is one called *Eyjafjallajökull*. The explosive phase of the eruption began on 14 April 2010. Winds blew the ash towards Europe and the airspace was closed down. Residents near airports such as London Heathrow heard birdsong in their gardens for the first time. While geologists, meteorologists and the International Civil Aviation Organization (ICAO) had taken the threat seriously, it seems the aviation industry had not – little if any research had been done, for instance, on how much ash an engine can tolerate. Airlines were up in arms over ICAO's precautionary stance. Ryanair's chief

executive said it was 'frankly ridiculous' that flight plans were being disrupted by 'an outdated, inappropriate and imaginary computer-generated model'. Even British Airway's boss was only somewhat less inflammatory in his criticism.

8   The most immense lava floods on Earth, such as the sixty-six-million-year old Deccan Traps of India, coincide with mass extinctions evident from the fossil record. There are many hypotheses for the 'kill mechanisms', but most incriminate the gases spewed into the atmosphere and then dissolved in the oceans. For a review, see: M. E. Clapham and P. R. Renne, 'Flood basalts and mass extinctions', *Annual Review of Earth and Planetary Sciences* 47 (2019): 275–303.

9   Nyiragongo in the Democratic Republic of Congo produces extremely fluid lavas that sometimes drain rapidly from a summit lava lake. Lava flows reportedly killed several hundred people in 1977, and several dozens of people were killed in a similar episode in 2002: J. C. Komorowski *et al.*, 'The January 2002 flank eruption of Nyiragongo Volcano (Democratic Republic of Congo): chronology, evidence for a tectonic rift trigger, and impact of lava flows on the city of Goma', *Acta Vulcanologica* 14 (2003): 27–62.

10   They may have affected air quality in northern and western Europe. I participated in a couple of exercises to evaluate the risk of Icelandic lava flood eruptions for the UK's National Risk Register. See: C. Witham *et al.*, 'UK hazards from a large Icelandic effusive eruption: Effusive Eruption Modelling Project final report', Met Office, 2015. https://core.ac.uk/display/33453938

11   C. E. Wieners, 'Haze, hunger, hesitation: disaster aid after the 1783 Laki eruption', *Journal of Volcanology and Geothermal Research* 406 (2020): 107080. K. Kleeman, *A Mist Connection: An Environmental History of the Laki Eruption of 1783 and its Legacy* (Berlin: De Gruyter, 2023).

12 Darwin's view was quite apocalyptic: 'At some future time contagious miasmata may be thus emitted from subterranean furnaces, in such abundance as to contaminate the whole atmosphere, and depopulate the earth!' While there were petitions for external aid to help the starving, there was no concerted scientific response to the eruption.

13 N. Mehler, 'The sulphur trade of Iceland from the Viking Age to the end of the Hanseatic period', *Nordic Middle Ages-Artefacts, Landscapes and Society. Essays in honour of Ingvild Øye on her 70th birthday*, UBAS-University of Bergen Archaeological Series 8 (Bergen: Universitetet i Bergen, 2015).

14 K. Poole, 'When hell freezes over: Mount Hecla and Hamlet's infernal geography', *Shakespeare Studies* 39 (2011): 152.

15 'Orkney – supposed volcanic eruption', *Dundee Courier*, 16 September 1845, Issue 1515.

16 As they did during further eruptions of Hekla in 1947 and 1970. Danish scientist Jørgen Christian Schythe joined the expedition of Bunsen and von Waltershausen in an independent capacity. His account of the 1845 eruption is foresighted in including sections on precursory phenomena (among them a reduction in milk yields!) and on the impacts of ashfalls on flora and fauna. Five years later (and fifteen years since Charles Darwin had gazed awestruck on its ruins) Schythe emigrated to Concepción in Chile. J. C. Schythe, *Hekla og dens sidste Udbrud den 2den September 1845* (Copenhagen: Trykt i Bianco Lunos Bogtrykkeri, 1847).

17 Beside its cartographic precision, it uniquely captured Etna's notable mutability. Inspired by Goethe's theory of colours, von Waltershausen represented the ages of the myriad lava flows enfolding the mountain in separate shades of green, a technique which remains the norm today (albeit without the fixation on a single colour). The map, however, has dated – it shows many features whose traces have since been swallowed up by more

than a century and a half of lava effusion and urban straggle.
S. Branca and T. Abate, 'Current knowledge of Etna's flank
eruptions (Italy) occurring over the past 2500 years. From the
iconographies of the XVII century to modern geological cartography', *Journal of Volcanology and Geothermal Research* 385
(2019): 159–178.

18   Bunsen's wider studies helped to build the foundations of spectroscopy that much of my approach to gas analysis has relied
on, and he laid groundwork for understanding sulphur chemistry. R. Bunsen, *Gasometry: comprising the leading physical and
chemical properties of gases* (London: Walton & Maberly, 1857).

19   He recovered and later reported that three-quarters of the
energy from charcoal fuel was wasted due to incomplete oxidation. Great efficiencies and cost-savings could be achieved, he
argued, if flue gases were reintroduced into the furnace. Had he
obtained a patent for the process, it would have made him a
great deal of money. At the time he was invited on the Icelandic
expedition, Bunsen was Professor of Chemistry at the University
of Marburg.

20   By the time the team reached Reykjavik, Hekla had finished
raging – an occupational hazard for a volcanologist. Volcanoes
don't wait for you to turn up.

21   This finding was reinforced by his analysis of samples from the
Beagle voyage, sent to him by Darwin, who was so impressed
with Bunsen's work that he persuaded Lyell to revise *Principles
of Geology* in light of it. Frank Perret was also a fan of Bunsen's
approach. R. Bunsen, 'Über den innern Zusammenhang der
pseudovulkanischen Erscheinungen Islands', *Justus Liebigs
Annalen der Chemie* 62 (1847): 1–59. For an excellent account of
Bunsen's work on Icelandic volcanism, see: C. Wentrup, 'Bunsen
the geochemist: Icelandic volcanism, geyser theory, and gas,
rock and mineral analyses', *Angewandte Chemie International
Edition* 60 (2021): 1,066–1,081.

Alexander von Humboldt drew on the volcano work of all our heroes – Oviedo, Hamilton, Spallanzani, Darwin, Scrope, Bunsen, von Waltershausen – in the fourth volume of his masterpiece, *Kosmos*, written late in his life. See: A. von Humboldt, '*Comsos: Sketch of the physical description of the universe*', trans. E. Sabine (London: Longman, 1858).

22 Yes – its nature has been experimentally determined as reported here, with attention paid to both smooth and crunchy varieties; G. P. Citerne *et al.*, 'Rheological properties of peanut butter', *Rheologica Acta* 40, (2001): 86–96.

23 F. Sauro *et al.*, 'Lava tubes on Earth, Moon and Mars: A review on their size and morphology revealed by comparative planetology', *Earth-Science Reviews* (2020): 103288.

24 See: A. B. Stefánsson and G. Stefánsdóttr, 'Surtshellir in Hallmundarhraun: Historical overview, exploration, memories, damage, an attempt to reconstruct its glorious past', *International Symposium on Vulcanospeleology* (2016).

25 Radiocarbon evidence suggests it initiated between around 880 and 910 CE. With an estimated volume of up to 8.5 cubic kilometres, it ranks as the third-largest eruption in Iceland since settlement.

26 For my money, the epitome of a domestic lava tunnel is the subterranean home of the Canarian artist and architect César Manrique (1919–1992) on Lanzarote.

27 N. Mari *et al.*, 'Potential futures in human habitation of martian lava tubes', in *Mars: A Volcanic World* (New York: Springer, 2021): 279–307.

28 K. P. Smith, 'Of monsters and men: literary, mythic, and archaeological views of Surtshellir Cave, Iceland', *Think UHI*, YouTube (uploaded 16 September 2016), https://www.youtube.com/watch?v=orRPQsHUEI8.

29 K. P. Smith *et al.*, 'Ritual responses to catastrophic volcanism in Viking Age Iceland: Reconsidering Surtshellir Cave through

Bayesian analyses of AMS dates, tephrochronology, and texts', *Journal of Archaeological Science* 126 (2021): 105316.

30 Orpiment was used as a pigment: the brightest yellow in illuminated manuscripts of the period.

31 M. Nordvig, *Volcanoes in Old Norse Mythology: Myth and Environment in Early Iceland* (Leeds: Arc Humanities Press, 2021).

32 Though the cave would have been accessible, no Viking-age artefacts were found deeper in the tunnel – as if there were a borderline that mortals should not cross.

33 An estimated 844 km² compared with 242 km² for Hallmundarhraun; see S. Sigurðardóttir *et al.*, 'Mapping of the Eldgjá lava flow on Mýrdalssandur with magnetic surveying', *Jökull* 65 (2015): 61–71.

34 C. Oppenheimer *et al.*, 'The Eldgjá eruption: timing, long-range impacts and influence on the Christianisation of Iceland', *Climatic Change* 147 (2018): 369–381.

35 Th. Thordarson *et al.*, 'New estimates of sulphur degassing and atmospheric mass-loading by the 934 AD Eldgjá eruption, Iceland', *Journal of Volcanology and Geothermal Research* 108 (2001): 33–54.

36 It is even possible to gauge the altitude of the dust from these medieval accounts. See Guillet *et al.*, 'Lunar eclipses illuminate timing and climate impact of medieval volcanism', *Nature*, 616 (2023): 90–95.

37 Trees can also record other stresses and disturbances, such as flood events, pest infestations and fires. It is an absolutely fascinating subject – for a primer, see: V. Trouet, *Tree Story: The History of the World Written in Rings* (Baltimore: John Hopkins University Press, 2020). Much of my recent work has been done in collaboration with Ulf Büntgen. His research has helped to show that, whereas individual eruptions might lead to a few years of cooling, clusters of eruptions in time can lead to much more prolonged cool periods that we see in 'little ice ages'. It

also looks like the warmer episodes of the past two millennia (excluding the recent warming period connected with fossil fuel use) coincide with intervals of statistically less volcanism. See: U. Büntgen *et al.*, 'Cooling and societal change during the Late Antique Little Ice Age from 536 to around 660 *AD*', *Nature Geoscience* 9 (2016): 231–236; U. Büntgen *et al.*, 'Prominent role of volcanism in Common Era climate variability and human history', *Dendrochronologia* 64 (2020): 125757.

38 L. M. Hollander (trans.), *The Poetic Edda* (Austin: University of Texas Press, 2nd edition, revised. Copyright © 1962, renewed 1990). Vǫluspá was already thought to have been first composed in the tenth century – if our interpretation is correct, then we can date it after 939 CE.

39 There were clear signs of the reawakening: uplift of the area was detected in the fall of 2020, followed in February 2021 by thousands of earthquakes signalling the surfacing of a sheet of magma, five miles long and a metre wide. Hundreds of tremors were felt in Reykjavik and, during the peak, more rattled the Reykjanes Peninsula in a single day than the average for a year. The eruption began with the opening of a 180-metre-long fissure at the surface.

40 Bunsen enjoyed enveloping his companions in fog by blowing cigar smoke into the steam vents. See Wentrup (ibid) and Bunsen's joyful letters home: R. E. Oesper and K. Freudenberg, 'Bunsen's trip to Iceland. As recounted in letters to his mother', *Journal of Chemical Education* 18 (1941): 253.

41 It put me in mind of Þríhnúkagígur (good luck with the pronunciation) in the east of the Reykjanes Peninsula, a huge, vault-like vertical lava tube. It has a narrow opening at the top of a cinder cone formed 3,500 years ago during a fissure eruption, similar to that of Fagradalsfjall. I visited it once – it is open to the public – descending to the floor in an open elevator.

42 G. B. M. Pedersen *et al.*, 'Volume, effusion rate, and lava

transport during the 2021 Fagradalsfjall eruption: Results from near real-time photogrammetric monitoring', *Geophysical Research Letters* (2022): e2021GL097125.

43 D. K. Chester *et al.*, *Mount Etna: the anatomy of a volcano*, (London: Chapman and Hall, 1985).

44 Some years ago, I met the director of Eldheimar, a 'museum of remembrance' on Heimaey. She was thirteen when the islanders' lives were inverted. 'I did not want to talk about the experience for twenty years – I felt so embarrassed to be a refugee,' she confided. 'I came back here after twenty years living in Germany. Like many, I did not find it amusing to dig up my story, to excavate memories.'

45 Haroun Tazieff, still smarting from his recent dismissal as director of volcanology in Paris, considered it a worthless effort: 'Persuaded by a somewhat inexperienced foreign volcanologist, Icelandic authorities agreed to use fire-boats to sprinkle water on a tongue of the thick lava flows … No arguments could prevent the exercise'. The 'inexperienced' volcanologist he refers to is Steve Sparks, who defended his position in print, branding Tazieff as subjective rather than informed. H. Tazieff, 'La Soufrière, volcanology and forecasting', *Nature* 269 (1977): 96–97. H. Sigurdsson and S. Sparks, 'What happened at Heimaey?', *Nature* 271 (1978): 108.

46 R. I. Tilling *et al.*, 'The Hawaiian Volcano Observatory: A natural laboratory for studying basaltic volcanism', US Geological Survey Professional Paper 1,801 (2014); J. P. Lockwood and A. Torgerson, 'Diversion of lava flows by aerial bombing – lessons from Mauna Loa volcano, Hawai'i', *Bulletin Volcanologique* 43 (1980): 727–741; J. Dvorak, *The Last Volcano: A man, a romance, and the quest to understand nature's most magnificent fury* (New York: Pegasus Books, 2015). T. A. Jaggar (1956, ibid).

47 F. R. Pulido and F. J. B. Delgado, 'La Palma: Un volcán sin nombre y sin olvido', *Norte de Salud Mental* 18, 66 (2022): 11–21.

Red Sea, Black Gold

1   J. W. Gregory, *The Great Rift Valley*, (London: John Murray, 1896): 324.

2   S. Sontag, *The Volcano Lover: A Romance*, (London: Random House, 1993).

3   See, for example: S. F. Hodgson, 'Obsidian: sacred glass from the California sky', *Geological Society, London, Special Publications* 273, (2007): 295–313.

4   Pliny the Elder, *The Natural History*, trans. J. Bostock and H. T. Riley (London: Henry G. Bohn, 1855): Book 36, Chapter 67. Available online at Perseus Digital Library, Tufts University. https://www.perseus.tufts.edu/hopper/text?doc=Perseus%3Ate xt%3A1999.02.0137%3Abook%3D36%3Achapter%3D67

5   I like the way archaeologist and skilled knapper John Sea describes stone tools as the 'residue of behavior': J. J. Shea, *Stone Tools in Human Evolution: Behavioral differences among techno-logical primates* (Cambridge: Cambridge University Press, 2016). The pioneers of geochemical provenancing were Colin Renfrew and Joe Cann, working in the mid-1960s. Colin said to me once: 'When you can identify precisely where the obsidian came from – if you can really pin that down – you are getting close to analysing the very origins of human mobility and the scale of it.' For a review, see: Y. V. Kuzmin *et al.*, 'Global perspectives on obsidian studies in archaeology', *Quaternary International* 542 (2020): 41–53.

6   M. Leakey, *Disclosing the Past: An autobiography* (New York: Doubleday, 1984).

7   Apparently, the tracks were first found by accident as two members of the team were playfully flinging elephant dung at each other (as you do). One jumped into a gully and almost stepped on the prints. Recent research reveals that another

hominin taxa also left tracks at Laetoli. See: R. L. Hay, and M. D. Leakey, 'The fossil footprints of Laetoli', *Scientific American* 246 (1982): 50–57; E. J. McNutt *et al.*, 'Footprint evidence of early hominin locomotor diversity at Laetoli, Tanzania', *Nature* 600, 468–471 (2021): 1–4; M. Leakey (ibid, 1984).

8   Her Ethiopian name is Dinkenesh. Her anatomy also indicated bipedal gait: D. C. Johanson, and M. Taieb, 'Plio-pleistocene hominid discoveries in Hadar, Ethiopia', *Nature* 260 (1976): 293–297.

9   After nitrogen and oxygen, argon is the atmosphere's third most abundant constituent. It is derived from decay of radioactive potassium.

10   The radiocarbon clock ticks much faster – the half-life of carbon-14 is 5,730 years.

11   R. C. Walter, 'Age of Lucy and the first family: single-crystal $^{40}Ar/^{39}Ar$ dating of the Denen Dora and lower Kada Hadar members of the Hadar Formation, Ethiopia', *Geology* 22 (1994): 6–10.

12   C. M. Vidal *et al.*, 'Age of the oldest known *Homo sapiens* from eastern Africa', *Nature* 601 (2022): 579–583.

13   G. C. P. King and G. N. Bailey, 'Dynamic landscapes and human evolution', *Geological Society of America Special Paper* 471 (2010): 1–19.

14   The dispersal history is uncertain but certainly complex. Some favour a southern route via the Bab el Mandab straits at the southern end of the Red Sea, others the Levant. See: A. Beyin, 'The Bab al Mandab vs the Nile-Levant: an appraisal of the two dispersal routes for early modern humans out of Africa', *African Archaeological Review* 23 (2006): 5–30. A. Beyin, 'The western periphery of the Red Sea as a hominin habitat and dispersal corridor: marginal or central?', *Journal of World Prehistory* 34 (2021): 279–316.

The southern route would have required swimming or rafting: J. Hill *et al.*, 'Sea-level change, palaeotidal modelling and hominin dispersals: The case of the southern Red Sea', *Quaternary Science Reviews* 293 (2022): 107719.

15 Gregory developed the ideas of the Austrian geologist, Eduard Suess, who had analysed surveys and rock samples collected in the 1880s by the Hungarian explorer Count Teleki. While Suess realised the Rift must result from the stretching of the Earth's crust in the direction perpendicular to the valley, it took Gregory's field observations to confirm that the movements on opposing parallel faults dropped the intervening strip of crust, forming the valley floor. Gregory further identified multiple generations of fault escarpments, with former lake beds now uplifted. The whole structure of the Rift, he argued, had to be geologically young.

Bailey Willis, an American seismologist, prominent in his day, appears to be one of the first to coin the term 'Gregory Rift'. See: B. Willis, *Studies in comparative seismology: East African plateaus and rift valleys*, Publication 470, (Washington: Carnegie Institution, 1936).

16 D. P. McKenzie *et al.*, 'Plate tectonics of the Red Sea and east Africa', *Nature* 226 (1970): 243–248; E. Bonatti *et al.*, 'The Red Sea: birth of an ocean', in *The Red Sea*, eds. N. Rasul and I. Stewart (Heidelberg: Springer Berlin, 2015), 29–44.

17 This is one of the world's youngest and best-preserved 'large igneous provinces'. Additionally, the region was stretched by doming of the plate. Fault lines developed separating blocks of the crust that then tilted like books collapsing on a shelf. This extension resulted in development of the great rifts on the African continent, but even more spectacularly, proceeded all the way to the separation of Africa and Arabia along the Red Sea and Gulf of Aden. As the crust thinned and subsided, volcanoes started sprouting along the seabed around five million

years ago, generating new ocean floor. Today, the Red Sea is 1,200 miles long and up to two miles deep. The ongoing divergence of Africa and Arabia is nearly two centimetres per year. In contrast, the dilation of the East African Rift is just a tenth as fast. It will be many millions of years before an ocean splits Ethiopia in half.

18 H. Tazieff, 'The Afar triangle', *Scientific American* 222 (1970): 32–41; F. Barberi *et al.*, 'Long-lived lava lakes of Erta 'Ale volcano', *Revue de Géographie Physique et de Géologie Dynamique* 15 (1973): 347–351.

19 H. Tazieff, *L'odeur du soufre: expédition en Afar* (Paris: Stock, 1975). H. Tazieff *at al.*, 'Tectonic significance of the Afar (or Danakil) depression', *Nature* 235 (1972): 144–147. For an overview of his scientific contribution in this period, see: J. H. Varet, 'H. Tazieff (1914-1998) Des années Afar au secrétariat d'État (1967-1986): la difficile mutation institutionnelle', *Travaux du Comité français d'Histoire de la Géologie* 3 (2009): 115–145.

20 As anticipated, all this kit was impounded in customs on arrival, and my immediate task was to extricate it.

21 Before reaching Asaita, Thesiger received a 'far from friendly' reception from a band of warriors, but references to a non-existent machine gun helped them to reach an understanding: W. Thesiger, 'The Awash river and the Aussa sultanate', *The Geographical Journal* 85 (1935): 1–19. See also: W. Thesiger, *The Danakil Diary: Journeys Through Abyssinia, 1930–34* (Toronto: HarperCollins Canada, 1996).

22 W. Munzinger, 'Journey across the Great Salt Desert from Hanfila to the foot of the Abyssinian Alps', *Proceedings of the Royal Geographical Society of London* 13 (1868): 219–224. The first European to hear of Erta 'Ale may have been the Irish-Basque geographer and polyglot Antoine D'Abbadie, who spent twelve years exploring Abyssinia, covering enormous

distances barefoot in local dress so as to blend in. He surveyed long tracts of the Red Sea coastline, compiled a 40,000-word dictionary of Amharic and mapped an area the size of Italy. While in Afar, he heard of '*une montagne qui fume toujours*'. The first European to climb Erta 'Ale was botanist Johannes Hildebrandt in 1873. His guides warned him of *jinn* that circled the summit on flying horses, and refused to accompany him, so he went alone. When he finally reached the top, the view into the chasm astounded and moved him: 'It is as if a pitch black sea churned up by a mighty hurricane has broken against the cliffs, forming towers of foam that run and twist and suddenly freeze stiff. Thus lies the deserted mass of rock like a tombstone to some great primeval force.' J. M. Hildebrandt, 'Erlebnisse auf einer Reise von Massûa in das Gebiet der Afer und nach Aden', *Z. Ges. Erdkunde Berlin* 10 (1875): 1–38.

23   Eight years later, I made it to Erta 'Ale with a French expedition led by Jean-Louis Cheminée (French for 'chimney', another great name for a volcanologist), a veteran of the Tazieff expedition.

In 2005, I joined an Ethiopian team to investigate the eruption of Dabbahu volcano – possibly the smallest volcanic episode I have ever come across! It had produced a veneer of ash, and a fuming spiny dome of lava had protruded from a great fissure in the ground. The local herders said several camels had been swallowed up when the ground opened. This event was part of an extended episode representing an increment in the widening of Afar. See: D. J. Ferguson *et al.*, 'Recent rift-related volcanism in Afar, Ethiopia', *Earth and Planetary Science Letters* 292 (2010): 409–418.

24   I assembled an archive of Landsat images going back to 1972, and declassified spy satellite photographs from the mid-1960s. This revealed when lava had overflowed the crater,

and established a timeline of heat output from the volcano. C. Oppenheimer and P. Francis, 'Remote sensing of heat, lava and fumarole emissions from Erta 'Ale volcano, Ethiopia', *International Journal of Remote Sensing* 18 (1997): 1,661–1,692.

25 Playfair's dispatches were communicated by his friend, Charles Beke, who had also travelled in Abyssinia. Playfair led a distinguished diplomatic service and was a prolific writer.

26 The authorities in Massawa, then in Ottoman hands, suspected French perfidy and dispatched messengers to discover the perpetrators. Four years later, Massawa was annexed by Egypt, and Werner Munzinger was the British consul there.

27 R. C. Walter *et al.*, 'Early human occupation of the Red Sea coast of Eritrea during the last interglacial', *Nature* 405 (2000): 65–69.

28 We were not able to estimate the sulphur output of the eruption, but did identify a modest summer cooling in tree-ring datasets. P. Wiart and C. Oppenheimer, 'Largest known historical eruption in Africa: Dubbi volcano, Eritrea, 1861', *Geology* 28 (2000): 291–294.

29 There was no volcano monitoring in Eritrea. There was not even a single seismometer operating in the whole country.

30 B. Goitom *et al.*, 'First recorded eruption of Nabro volcano, Eritrea, 2011', *Bulletin of Volcanology* 77 (2015): 1–21.

31 Little is published on the region's volcanic geology, though there are reports of an eruption as recently as 1937.

32 This suggested distinct spheres of interaction on the highlands and lowlands. Lamya was struck by how little evidence there was for Neolithic exploitation of obsidian outcrops on Jebel Isbil volcano. We dated an explosive eruption to 6,000 years ago. Perhaps this had kept Neolithic people away. L. Khalidi *et al.*, 'Obsidian sources in highland Yemen and their relevance to archaeological research in the Red Sea region', *Journal of Archaeological Science* 37 (2010): 2,332–2,345.

33 C. Oppenheimer *et al.*, 'Risk and reward: explosive eruptions and obsidian lithic resource at Nabro volcano (Eritrea)', *Quaternary Science Reviews* 226 (2019): 105995.

34 L. Khalidi *et al.*, 'Obsidian use and interaction across Arabia and the Horn of Africa during the Holocene', in *Sourcing Obsidian*, eds. F. X. Le Bourdonnec *et al.* (New York: Springer 2023).

35 N. Blegen, 'The earliest long-distance obsidian transport: Evidence from the ~200 ka Middle Stone Age Sibilo School Road Site, Baringo, Kenya', *Journal of Human Evolution* 103 (2017): 1–19.

36 E. Yohannes, 'Geothermal exploration in Eritrea—country update', *Proceedings, World Geothermal Congress,* 2015, Melbourne, Australia, International Geothermal Association (2015).

There is a vast potential geothermal resource in the Rift Valley and Afar. See: N. Burnside *et al.*, 'Geothermal energy resources in Ethiopia: Status review and insights from hydrochemistry of surface and groundwaters', *Wiley Interdisciplinary Reviews:Water*, 8(6), e1554 (2021).

Water Tower of the Sahara

1 G. Nachtigal, *Sahara et Soudan*, vol. 1, trans. J. Gourdault (1881): 176.

2 W. Thesiger, 'A camel journey to Tibesti', *The Geographical Journal* 94, (1939): 433–446.

3 Radar instruments carried in space operate by the same basic principle, but look sideways and build two-dimensional images with very fine-scale detail. The acuity of the radar depends in part on the dimensions of its antenna – the longer it is, the more detailed the imagery. From the great distance of Earth orbit, an antenna might need to be miles long to resolve small features at the surface. This is impractical but, amazingly, such an

antenna can be simulated with a very modest aerial thanks to a satellite's motion, in effect by combining multiple consecutive images, each of which has been acquired a bit further along in the orbit.

4   For a great overview of Magellan's discoveries of Venusian volcanoes, see: J. W. Head *et al.*, 'Venus volcanism: Classification of volcanic features and structures, associations, and global distribution from Magellan data', *Journal of Geophysical Research: Planets* 97 (1992): 13,153–13,197. There is evidence that volcanic activity persists today on Venus. See: R. H. Herrick and S. Hensley, 'Surface changes observed on a Venusian volcano during the Magellan mission', *Science* (2023): doi:10.1126/science.abm7735.

5   French geologists reconnoitred the volcanoes in the 1930s and then again in the 1950s. The most recent work, carried out in the 1970s, was based on interpretations of satellite images. It compared the massive shield and caldera of Emi Koussi with the Elysium volcanoes on Mars. The scientist behind this work, Mike Malin, went on to design the camera that provided stunning images of Mars and revealed the former presence of water. Today, it is safe to say we know more about the geology of the Red Planet than of the Tibesti. See: M. C. Malin, 'Comparison of volcanic features of Elysium (Mars) and Tibesti (Earth)', *Geological Society of America Bulletin* 88 (1977): 908–919.

6   At the time, I was supervising a PhD student, Jason Permenter, who was building sophisticated computer-animated fly-bys of the Tibesti's volcanoes using all the best space imagery and digital topography we could lay our hands on. His studies might have been revolutionary but for the unanticipated launch of Google Earth. See: J. L. Permenter and C. Oppenheimer, 'Volcanoes of the Tibesti massif (Chad, northern Africa)', *Bulletin of Volcanology* 69, (2007): 609–626.

7 After wartime service training pilots to fly one of the Royal Air Force's fastest bombers in Canada, Dick returned to Cambridge to complete a degree in Geography. He studied under Frank Debenham, a veteran of Captain Scott's final Antarctic campaign. Between stories of polar adventure, Debenham lectured on his current interest, water resources in Africa. It was through Debenham's connections that Dick was invited a few years later to join a mission to the Sahara. For a fascinating recorded conversation with Dick Grove, see: 'Dick Grove', Voices of Science, British Library, https://www.bl.uk/voices-of-science/interviewees/dick-grove.

8 Q. Schiermeier, 'Exposing Sahara science in the shadow of terrorism', *Nature News* (2015).

9 A. T. Grove and A. Warren, 'Quaternary landforms and climate on the south side of the Sahara', *The Geographical Journal* 134 (1968): 194–208.

10 R. Kuper and S. Kröpelin, 'Climate-controlled Holocene occupation in the Sahara: motor of Africa's evolution', *Science* 313 (2006): 803–807.

11 Hotter Northern Hemisphere summers drove moisture off the Atlantic Ocean, intensifying the western African monsoon and bringing rain further north. There have been multiple earlier humid periods, including one around 125,000 years ago, which established a vegetated corridor across northern Africa, facilitating migration of both archaic and modern humans as far as the Mediterranean coast – and potentially, into Europe. The Tibesti, in particular, provided a refuge for flora and fauna, including humans, during hyper-arid phases, while also acting as a 'centre of propagation' when the Sahara greened. On the orbital mechanics that influence climate of northern Africa, see: L. Menviel *et al.*, 'Drivers of the evolution and amplitude of African Humid Periods', *Communications Earth & Environment* 2, no. 237 (2021). On the multiplicity of humid periods, see:

J. Larrasoaña *et al.*, 'Dynamics of green Sahara periods and their role in hominin evolution', *PloS One* 8 (2013): e76514.

12 The conventional view sees the stimulus arriving from southwest Asia. Want to experience Stefan Kröpelin for yourself? There is a great presentation he gave for the Long Now Foundation that is available online: S. Kröpelin, 'Civilization's Mysterious Desert Cradle: Rediscovering the Deep Sahara', the Long Now Foundation, https://longnow.org/seminars/02014/jun/10/civilizations-mysterious-desert-cradle-rediscovering-deep-sahara/.

13 J. Brachet and J. Scheele, 'Remoteness is power: disconnection as a relation in northern Chad', *Social Anthropology* 27 (2019): 156–171.

14 W. Thesiger, 'A camel journey to Tibesti', *The Geographical Journal* 94 (1939): 433–446.

15 In 2011, $60 million USD of funding was pledged by the central government for development projects in the Tibesti, but violence and lawlessness, combined with embezzlement and weak management, led to the collapse of the programme.

16 Together, they steered successful proposals for both of Chad's UNESCO World Heritage Sites – the Ennedi Plateau and Ounianga lakes.

17 Faya was the base for the geographer and military commander Jean Tilho (1875–1956), who explored the Tibesti in 1916. At this time, the Sanūssiyya, a Sufi movement fighting colonial expansion in northern Africa, controlled the mountains. Tilho had captured Faya in 1913, probably playing a decisive role in shielding French and British protectorates south of the Sahara from German and Turkish forces during the First World War. He later set off with a detachment of men and camels on a ten-week expedition in the Tibesti, with the aim of wresting control from the Sanūssiyya. Their first adventure was the ascent of Emi Koussi. The scene of 'wild magnificence' seen from the summit stayed with Tilho for the rest of his life. See: J. Tilho, 'The

exploration of Tibesti, Erdi, Borkou, and Ennedi in 1912–1917: A Mission entrusted to the author by the French Institute (Continued)', *The Geographical Journal* 56 (1920): 161–183.

18 R. M. Prothero, 'Heinrich Barth and the western Sudan', *The Geographical Journal* 124 (1958): 326–337.

19 G. Nachtigal, *Sahara and Sudan. Vol. I: Tripoli and Fezzan, Tibesti or Tu*, trans. A. G. B. Fisher and H. J. Fisher (London: C. Hurst and Co, 1974). See also H. Fisher, 'Dr. Gustav Nachtigal in Modern Chad', *Itinerario*, 9 (1985): 145–156, in which the author presciently notes that the Tibesti 'is likely to prove a singularly ill-fitting piece for the mosaic of any wider national state'.

20 A. T. Grove, 'Geomorphology of the Tibesti region with special reference to western Tibesti', *The Geographical Journal* 126 (1960): 18–27.

21 As we prepared to break camp, a group of Chadians pulled up in a Hilux. I noticed a couple of metal detectors in the back. They showed us a sack full of spectacular meteorites they had found in the mountains, curious to know what we thought they were worth.

22 Ottoman administration came briefly to the Tibesti in the early twentieth century, somewhat in alliance with the Sanūssiyya. See: G. Joffé, 'Chad: Power vacuum or geopolitical focus', *Geopolitics* 2 (1997): 25–39.

23 For more on the concept of mountains as water towers, see: W. W. Immerzeel *et al.*, 'Importance and vulnerability of the world's water towers', *Nature* 577 (2020): 364–369.

24 C. Deniel *et al.* 'The Cenozoic volcanic province of Tibesti (Sahara of Chad): Major units, chronology, and structural features', *Bulletin of Volcanology* 77 (2015): 1–21.

25 P. W. Ball *et al.*, 'Quantifying asthenospheric and lithospheric controls on mafic magmatism across North Africa', *Geochemistry, Geophysics, Geosystems* 20 (2019): 3,520–3,555.

26   A range of studies of the diatomites have yielded new insights
into the hydroclimate of the Sahara: A. N. Yacoub *et al.*, 'The
African Holocene Humid Period in the Tibesti mountains
(central Sahara, Chad): Climate reconstruction inferred from
fossil diatoms and their oxygen isotope composition', *Quaternary
Science Reviews* 308 (2023): 108099.

27   Large calderas are formed by subsidence of the ground above
a magma reservoir when a large quantity of the molten rock is
expelled, either by eruption or via dikes that squeeze magma
laterally into the surrounding crust, potentially over great
distances (this occurred at Miyakejima volcano in Japan in
2000). There is no evidence for a very large eruption of Trou
au Natron in the past millennia, so the remaining question is
whether dikes could have withdrawn magma from beneath
the volcano, triggering deepening of the existing caldera. But
looking at the topography of the caldera floor, this is as
unlikely a scenario for me as a 500-metre-deep lake is for
Stefan.

Flame in a Sea of Gold

1   'At the edge of the crater', by George Marston in *Aurora
Australis* (Cape Royds: British Antarctic Expedition, 1908).
2   J. O'Reilly, 'Sensing the ice: field science, models, and expert
intimacy with knowledge', *Journal of the Royal Anthropological
Institute* 22 (2016): 27–45.
3   In Aristophanes' 414 BCE play *Birds*: 'At the beginning there
was only Chaos, Night, dark Erebus, and deep Tartarus. Earth,
the air and heaven had no existence. Firstly, black-winged Night
laid a germless egg in the bosom of the infinite deeps of Erebus,
and from this, after the revolution of long ages, sprang the
graceful Eros with his glittering golden wings, swift as the whirlwinds of the tempest.' Aristophanes, *Birds*, in *The Complete*

Greek Drama, vol. 2., W. J. Oates and E. O'Neill, Jr. (New York: Random House, 1938). Available online at Perseus Digital Library, Tufts University. https://www.perseus.tufts.edu/hopper/text?doc=Perseus%3Atext%3A1999.01.0026%3Acard%3D685

4 R. Yeo, *Defining Science: William Whewell, natural knowledge and public debate in early Victorian Britain* (Cambridge: Cambridge University Press, 2003).

5 F. Fleming, *Barrow's Boys* (New York: Grove Press, 2001).

6 Bomb vessels were the weapons of mass destruction of their day – their conversion for science was a real case of 'swords to ploughshares'. See: N. Robins, *From War to Peace: The conversion of naval vessels after two world wars* (Barnsley: Seaforth Publishing, 2021); A. W. H. Pearsall, 'Bomb vessels', *Polar Record* 16 (1973): 781–788.

7 Though by this time, he was a Member of Parliament and writing on economics as much as on geology.

8 The mineralogist George Prior examined the rocks collected by McCormick, whose 'so-called geological accounts' he found, in most cases, to 'resolve themselves into exasperating (from a petrological point of view) descriptions of birds, for the doctor appears to have been a more enthusiastic ornithologist than geologist'. G. T. Prior, 'Petrographical notes on the rock specimens collected in Antarctic regions during the voyage of HMS *Erebus* and *Terror* under Sir James Clark Ross, in 1839–43', *Mineralogical Magazine* 12 (1899): 69–91. See also: M. J. Ross, *Ross in the Antarctic: the Voyages of James Clark Ross in Her Majesty's Ships Erebus & Terror, 1839–43* (Whitby: Caedmon of Whitby, 1982); J. C. Ross, *A Voyage of Discovery and Research in the Southern and Antarctic Regions, during the Years 1839–43*, vol. 1 (London: John Murray, 1847).

9 Thank goodness Ross had not sailed to Antarctica in some other naval ships of the era that were associated with polar exploration, such as HMS *Racehorse* or HMS *Carcass*. An astonishing testament to the accuracy of naval surveying is that Ross and his

crew estimated Mount Erebus to be 12,400 feet (3,780 metres) high, which is almost spot on. Reflecting on how far off they were at sea on a lurching sail ship, it really is remarkable.

10   I, too, dreamed of over-wintering on Erebus – my fieldwork, though, was only ever possible in the austral summer, when the sun does not set. How often I imagined the spectacle of the frosted cone of Erebus touching a night sky of scintillating star-light and glaucous auroral phosphorescence, steam billowing from the crater diffusing the bright red glow of the lava lake.

11   R. F. Scott, *The Voyage of the Discovery*, vol. 1 (London: Smith, Elder & Co, 1907).

12   Knowledge of Mount Erebus swiftly permeated the scientific literature of the day. Hooker beat Ross to press with *Botany of the Antarctic Voyages* in 1844, followed by Ross's account, published in two volumes in 1847. Humboldt mentions the volcano's discovery in a section on terrestrial magnetism in the 1854 edition of *Kosmos*. And at the first opportunity, the volcano is introduced in Lyell's *Principles of Geology* (in its seventh edition, 1847). Lyell mentions Mount Erebus in a section on polar climate, noting that all of Victoria Land above 4,000 feet was covered in perpetual snow, 'except a narrow ring of black earth surrounding the huge crater of the active volcano'. Similarly, the second edition of George Poulett Scrope's classic work on volcanoes draws on Ross's observations, though some-what embellishes them. J. D. Hooker, *The Botany of the Antarctic voyage of HM Discovery Ships* Erebus *and* Terror *in the Years 1839–1843: Under the command of Captain Sir James Clark Ross*, vol. 1 (London: Reeve Brothers, 1844).

13   Edgeworth David was one of Professor John Wesley Judd's students at the Royal School of Mines in London. Judd, in turn, was mentored by George Poulett Scrope (who, when he was no longer active enough to do his own fieldwork, commissioned the younger man to survey the volcanoes of Europe). Judd helped

to establish the study of volcanic rocks under the microscope, and wrote a popular book on volcanoes, as well as the geological sections of the Royal Society's Krakatoa report of 1888. He counted Charles Lyell and Charles Darwin among his friends.

My collaborators at University of Chicago Press would have liked a chapter or two on volcanoes of the United States in *Mountains of Fire*. I've loved working on Kīlauea with colleagues at the Hawaiian Volcano Observatory but my experiences there didn't really sum up to a whole chapter's worth of adventure. I have also seen Mount St Helens, Crater Lake, Mammoth Mountain, the Inyo volcanic chain, Yellowstone and the Valles caldera of the Jemez mountains – all magnificent and instructive but these encounters were geotouristic and not dedicated field missions. I hope, despite the thin coverage on the nation's immense volcanic heritage, to avoid a review like the one the American geologist, Alexis Julien, penned for Judd's volcano book: 'The little hills of England, Wales and Scotland afford excellent illustrations of extinct volcanic vents for the British public; but why, O why, should not discreet reference be vouchsafed to transatlantic localities for the benefit of Prof. Judd's thirty odd millions of cousins?' Julien continued: '[There are] eighty-two great volcanic vents in action on our continents, to only thirty-five in Europe, Asia and Africa ... this fact ... indicates that America is the grandest field in the world for the study of volcanic phenomena; and this little book, excellent as it is, with the limited field of observation from which it has been mainly written, gives a suggestion of the magnificent monograph on this subject which is yet to come from some American hand.' J. W. Judd, *Volcanoes: What they are and what they teach*, (New York: Appleton and co., 1881). Julien's review is in: *Science* 77 (1881): 600–601.

14 Carsten Borchgrevink, a Norwegian in command of the *Southern Cross* Expedition, was first to land on Ross Island,

eight years earlier, at the foot of Mount Terror. He was speaking of his exploits in New York in 1902 as a guest of the National Geographic Society. It was shortly before the Montagne Pelée disaster, following which the Society funded a scientific team to investigate. Probably exaggerating his credentials as a volcanologist, Borchgrevink was invited to join, along with geologists Israel Russell and Bob Hill. Thomas Jaggar, who would found the Hawaiian Volcano Observatory a decade later, and Edmonds Hovey, from the American Museum of Natural History, also travelled to Martinique. They arrived there on 21 May, a day after another violent eruption of Pelée had killed 2,000 people. Back in New York, Borchgrevink made numerous exaggerated claims that moved Hovey to denounce him in print so as to scupper the Norwegian's chances of fundraising for another Antarctic expedition. It worked.

Among the last to know of the Pelée disaster was Shackleton, who was overwintering on Ross Island with Captain Scott's *Discovery* Expedition of 1901–1904. John Gregory (of East African Rift fame) was slated to be scientific director of that expedition, but refused to go there under naval command and resigned.

15 Shackleton was clearly charmed by David. In a letter to his wife, he described the Professor as: 'a powerful intellect . . . a charming companion and a tower of strength . . . he could charm a bird off a bough'. M. E. David, '*Professor David: The life of Sir Edgeworth David*' (London: Edward Arnold, 1937).

16 David knew his Shakespeare; conjuring up *A Midsummer Night's Dream* on the side of a frozen volcano is inspired.

17 B. M. Tebo *et al.*, 'Microbial communities in dark oligotrophic volcanic ice cave ecosystems of Mt. Erebus, Antarctica', *Frontiers in Microbiology* 6 (2015): 179; C. Coleine and M. Delgado-Baquerizo, 'Unearthing terrestrial extreme microbiomes for searching terrestrial-like life in the Solar System', *Trends in Microbiology* 30 (2022): 1,101–1,115.

18 David, Mawson and Mackay were reunited as the party that
    located (as much as a moving target can be) the south magnetic
    pole, an exceptionally arduous and unaided trek of over 1,200
    miles. David, who suffered terribly from exhaustion on the
    return journey, later described Mawson as 'the real leader who
    was the soul of our expedition'. Mawson was back in Antarctica
    in 1911, leading the Australasian Antarctic Expedition. He was
    the lone survivor of an overland trek during which his two
    colleagues died. He later returned to the University of Adelaide
    as Professor of Geology and Mineralogy. He died in 1958.
    Mackay's life was cut short in 1914, somewhere near Wrangel
    Island in the Siberian Sea, during a failed attempt to reach land
    following the sinking in pack ice of the Canadian Arctic
    Expedition's brigantine *Karluk*. David, a man of 'artistic
    temperament allied to scientific training', died in 1934 and was
    given a state funeral. See: R. E. Priestley, 'Sir Edgeworth David',
    The Australian Quarterly 10 (1938): 34–39.

    See also: E. J. Larson, *An Empire of Ice: Scott, Shackleton, and
    the Heroic Age of Antarctic Science* (New Haven: Yale University
    Press, 2011); F. Spufford, *I May Be Some Time: Ice and the
    English Imagination* (London: Faber & Faber, 1996); E. J.
    Larson, 'Public science for a global empire: The British quest
    for the south magnetic pole', *Isis* 102 (2011): 34–59.

19 T. W. E. David and R. E. Priestley, 'Glaciology, physiography,
    stratigraphy, and tectonic geology of south Victoria Land', in
    *Reports on the Scientific Investigations, British Antarctic Expedition
    1907–9*, vol. 1 (London, William Heinemann: 1914).

20 E. J. Larson (ibid).

21 Someone else who leapt for joy on receipt of a similar invitation
    from Phil, but three decades earlier, was Haroun Tazieff, who
    wrote: '. . . *je reçus une lettre qui me laissa pantois: revenant de
    l'Erebus dont il avait avec ses compagnons tenté de descendre le
    cratère, le jeune géologue néo-zélandais Philip Kyle m'invitait à*

prendre l'année suivante la tête d'une nouvelle tentative! Je suis incapable de décrire l'espèce de cyclone qui a déclenchérent en moi la lecture de ce message, sa relecture à la recherche de quelque indice de canular possible, puis la lente, la progressive compréhension que se réalisait le rêve entretenu durant tant d'années.' H. Tazieff, *Erebus: volcan Antarctique* (Paris: Arthaud, 1978).

22   The closest I came to it was one day at Lower Erebus Hut – I'd made myself a coffee and thought, since I'd made a couple of strenuous trips up to the crater, that I'd add a spoonful of sweetened condensed milk to it. There was a tin of the stuff sitting by the cooker. It really didn't taste great but then we've got cans and packets of food up there whose 'use by' dates stretch back almost to Scott's era. It turned out it was bacon fat.

23   *Encounters at the End of the World* (2007), Werner Herzog's Oscar-nominated movie, gives a flavour of the oddities populating MacTown. It was atop Erebus that I met Herzog for the first time, sparking our friendship and film-making collaboration.

24   There was a whiteboard in our field hut on Erebus on which we recorded any items we needed from McMurdo – food resupplies, snowmobile parts, another U-barrel (for urine – all human waste was collected). We'd call in with the list and the goods would arrive the next time the helicopter could fly in. I laughed out loud one time (and thought of the Scott and Shackleton era) when I saw on the whiteboard someone had written beside a request for 'Freshies': 'Could we *not* have Lollo Rosso this time?'

25   That being said, I learned during my first year with the U.S. Antarctic Program that the coffee provided on an American base is undrinkable – dispensed from a large and perpetually brewing cauldron, it mostly tasted of rubber gaskets. Thereafter, coffee was the one foodstuff I insisted on bringing from home. My last mission to Antarctica was to South Korea's Jang Bogo station, close to Mount Melbourne, another lofty frozen volcano.

I loved it on-base – it was so quiet, with just fifty residents, and there was a professional barista coffee machine.

26   Pee flags were phased out to protect the environment in my later field seasons with the U.S. Antarctic Program. All human waste had to be contained and shipped off the continent. Some say it is processed into fertiliser and biofuel in the U.S. Whatever happens, the barrels have to cross the equator on that voyage, and I wouldn't want to walk past them aboard the ship.

27   We all noticed the altitude at Fang – just moving gear from the helicopter to the tents had me gasping for air.

Some took Diamox to prevent acute mountain sickness. I tried it once and can vouch for its diuretic action. It was quite unwelcome in my case as I was in my sleeping bag and using a 'pee bottle' (standard equipment for all in field parties, issued in McMurdo). I filled it at such an alarming rate that there was no time to pull back. After that, my pals thoughtfully made up a mini 'spill kit' for me, complete with a funnel, extra bottle and a folding containment berm.

28   D. Sweeney *et al.*, 'Sulfur dioxide emissions and degassing behavior of Erebus volcano, Antarctica', *Journal of Volcanology and Geothermal Research* 177 (2008): 725–733.

29   P. J. Fraser, 'Global and Antarctic ozone depletion: A rebuttal of the critics', *Clean Air: Journal of the Clean Air Society of Australia and New Zealand* 24, no. 4 (1990): 139–143. C. Oppenheimer, 'Atmospheric chemistry of an Antarctic volcanic plume', *Journal of Geophysical Research*, 115, D04303 (2010): doi:10.1029/2009JD011910.

30   In the 1970s, Phil had worked on Erebus with a brilliant German geochemist, Werner Giggenbach. His volcanic gas sampling apparatus became known as the Giggenbach bottle. I have lived by his creed: 'Nature only reveals her secrets if we ask the right questions and listen, and listening in geochemistry means sampling, analyzing, plotting'. See: W. F. Giggenbach, 'Magma

degassing and mineral deposition in hydrothermal systems along convergent plate boundaries', *Economic Geology* 87 (1992): 1,927–1,944.

31 Courrejolles was an engineer living in Hispaniola. He lost a mass of unpublished manuscripts, 'the fruit of thirty years labour in the arts and sciences' during the Haitian revolution. 'Nothing but experience, observation, and comparison ... can conduct us with wisdom in our researches,' he wrote. The title of his paper is inspired: F. Courrejolles, 'Observations which seem to prove the Necessity of observing and meditating a long time before any decisive opinion is formed in Philosophy in general, and particularly in regard to the cause of earthquakes', *The Philosophical Magazine* 12 (1802): 337–346.

32 C. Oppenheimer *et al.*, 'Pulsatory magma supply to a phonolite lava lake', *Earth and Planetary Science Letters* 284 (2009): 392–398.

33 You might wonder why we couldn't use a laser ranger for this – we tried, but we just couldn't get a return off the lava surface due to the fumes. See: N. J. Peters *et al.*, 'Radar altimetry as a robust tool for monitoring the active lava lake at Erebus volcano, Antarctica', *Geophysical Research Letters* 45 (2018): 8,897–8,904.

34 The explanation developed from the observations I was involved in of cyclic lava fountains at Fagradalsfjall, Iceland, in 2021. See: S. Scott *et al.*, 'Near-surface magma flow instability drives cyclic lava fountaining at Fagradalsfjall, Iceland', *Nature Communications* in review.

35 I like how the Irish geologist, Edward Hull, put it in his volcano book: 'Macroscopic and microscopic observations have to go hand in hand in the study of volcanic phenomena'. E. Hull, *Volcanoes: past and present*, (London: Walter Scott, 1892).

36 My studies took me to the labs at Arizona State University to initiate experiments on Erebus rocks with Kayla Iacovino. The

aim was to figure out how much water and carbon dioxide the parent magmas beneath the volcano could hold. I'd spoon minute quantities of powdered lava, water and oxalic acid (which yields CO_2 on heating) into miniscule gold capsules. Their lids had then to be sealed using a spot welder. A leaky join would ruin the experiment, as would overcooking the sample under the blinding arc. It took a while for me to find the dexterity needed to manipulate the electrode tip, seen through a binocular microscope. The tiny cans were then loaded individually into an apparatus called a piston cylinder capable of imposing phenomenal pressures and cooking at magmatic temperatures. Experiments were carried out at different temperatures and pressures to determine how much water and CO_2 dissolved into the molten rock in each case. K. Iacovino *et al.*, 'H_2O–CO_2 solubility in mafic alkaline magma: applications to volatile sources and degassing behavior at Erebus volcano, Antarctica', *Contributions to Mineralogy and Petrology* 166 (2013): 845–860.

37 Y. Moussallam *et al.*, 'Tracking the changing oxidation state of Erebus magmas, from mantle to surface, driven by magma ascent and degassing', *Earth and Planetary Science Letters* 393 (2014): 200–209.

38 C. Oppenheimer *et al.*, 'Mantle to surface degassing of alkalic magmas at Erebus volcano, Antarctica', *Earth and Planetary Science Letters* 306 (2011): 261–271.

39 Y. Moussallam *et al.*, 'Experimental phase-equilibrium constraints on the phonolite magmatic system of Erebus volcano, Antarctica', *Journal of Petrology* 54 (2013): 1,285–1,307; Y. Moussallam *et al.*, 'Megacrystals track magma convection between reservoir and surface', *Earth and Planetary Science Letters* 413 (2015): 1–12.

40 I had speculated on this process with Peter Francis a long time ago. P. Francis *et al.*, 'Endogenous growth of persistently active volcanoes', *Nature* 366 (1993): 554–557.

41 For example: D. Stewart, 'Petrography of some erratics from Cape Royds, Ross Island, Antarctica', *American Mineralogist* 44 (1959): 1,159–1,168.

42 Bar Dickason and Debenham, who were suffering from altitude sickness, and so remained at the Fang Glacier mapping and geologising.

    Gran built a cairn on the crater rim in which they would leave a time capsule, and then he and Priestley started taking photographs. On the way back down, the men realised they had accidentally put a tin containing exposed films in the cairn. Gran went to retrieve it. Priestley was just approaching the base of the summit cone when there was an ear-splitting boom. Turning on his heels, he saw lava bombs whiz through the air then fizz in the snow. Gran was not to be seen. Priestley raced back up fearing the worst. He had just reached the crater lip when a nonplussed but essentially unimpaired Gran appeared from a cloud of steam. The Norwegian later said of the experience that it was 'disagreeable in the extreme'.

43 The site, along with another that I located below the Fang Glacier camp, is now inscribed as a Historic Site and Monument, and protected under the Antarctic Treaty.

44 F. Debenham, 'The future of polar exploration', *The Geographical Journal* 57 (1921): 182-200.

45 We titled our research project 'Mount Erebus Volcano Observatory', another variation on the observatory theme.

46 In 'the first book ever written, printed, illustrated and bound in the Antarctic': E. H. Shackleton *et al.*, *Aurora Australis*, 'printed at the sign of "the penguins"'. See note 1 of this chapter.

The Volcano and You

1 Artwork by W. Simpson, 'Sketches of Mount Vesuvius', *The Illustrated London News* 60, No. 1706 (1872): 479-481.

2 This was from a lovely card I received, decorated with Lucy's painting of an erupting volcano.

3 Around 1,350 active volcanoes are documented. S. Freire *et al.*, 'An improved global analysis of population distribution in proximity to active volcanoes, 1975–2015', *ISPRS International Journal of Geo-Information* 8, 341. (2019): doi:10.3390/ijgi8080341

4 Admittedly, Baiæ also had a reputation for raucous beach parties and amorous encounters, and the lure of debauchery was perhaps as strong as that of therapy. L. Giacomelli and R. Scandone, 'History of the exploitation of thermo-mineral resources in Campi Flegrei and Ischia, Italy', *Journal of Volcanology and Geothermal Research* 209 (2012): 19–32. A. Costa *et al.*, 'The long and intertwined record of humans and the Campi Flegrei volcano (Italy)', *Bulletin of Volcanology* 84 (2022): 1–27.

   Of course, we enjoyed our balneological experiences so much, we invented the bathroom – if you can't get to the volcano, bring the volcano into your home.

5 *The Illustrated London News* 60, No. 1706 (1872): 479–481.

6 F. V. Hayden, 'The Yellowstone National Park', *American Journal of Science and Arts* 3 (1872): 294–297. R. Nash, 'The American invention of national parks', *American Quarterly* 22, (1970): 726–735.

7 Attracting visitors to potentially dangerous volcanic areas demands that effective risk management operations are in place. There have been tragic incidents on active volcanoes, including the 2019 disaster at Whakaari (White Island), New Zealand, which claimed the lives of 22 tourists and guides. See: P. Erfurt, 'Volcano Tourism and Visitor Safety: Still Playing with Fire? A 10-Year Update', *Geoheritage* 14, 56 (2022).

8 T. J. Casadevall *et al.*, 'Protecting our global volcanic estate: review of international conservation efforts', *International Journal of Geoheritage and Parks* 7 (2019): 182–191.

9   E. Pijet-Migoń and P. Migoń, 'Geoheritage and cultural heritage—A review of recurrent and interlinked themes', *Geosciences* 12(2) 98 (2022): doi:10.3390/geosciences12020098. T. A. Semeniuk, 'The Hornsby Quarry geosite, NSW, Australia—A geoheritage treasure', *Land* 11, 2124 (2022).

10 A nomination from Pyongyang for Mount Paektu is under consideration. The volcano's biodiversity is already recognised with UNESCO Biosphere Reserves designated on each side of the frontier running through the crater.

11 In tandem with the geoparks movement, there are museums, exhibitions, festivals, films and excursion guides that seek to elevate our understanding of volcanoes. There's even a proposal to designate the 10th of April – the day of Tambora's great convulsion – as an 'international day of volcanoes'. And here are my top ten volcano films: *Stromboli* (1949/50), *La Soufrière* (1977), *Dante's Peak* (1997), *El Valle sin Sombras* (2015), *Ascent* (2016), *Into the Inferno* (2016), *Volcano: what does a lake dream?* (2019), *Merapi* (2021), *The Fire Within* (2022), *Fire of Love* (2022).

12 I know, you're thinking 'but wasn't that the asteroid's fault?'. It is quite a conundrum – the formation of the Chicxulub impact crater in Yucatan, México, *and* the peak volcanism of the Deccan Traps, *and* the mass extinction all happen at about the same time, 66 million years ago. But there are several extinction events recognized from the fossil record of the past few hundred million years, and *all* coincide with massive lava floods, implicating the associated gas emissions and climatic impacts of the eruptions as the primary cause. In the case of the Deccan, it is even speculated that the impact played a role in triggering the volcanism. See: T. Mittal *et al.*, 'Deccan volcanism at K-Pg time', *Geological Society of America Special Paper* 557 (2022): 471–496

13 The Elephanta caves are also inscribed on the World Heritage List. H. Sheth *et al.*, 'The volcanic geoheritage of the Elephanta

caves, Deccan traps, Western India', *Geoheritage* 9 (2017): 359–372. G. Kaur *et al.*, 'The late Cretaceous-Paleogene Deccan traps: a potential global heritage stone province from India', *Geoheritage* 11 (2019): 973–989.

14 On the bathing monkeys, see: R. Takeshita *et al.*, 'Beneficial effect of hot spring bathing on stress levels in Japanese macaques', *Primates* 59 (2018): 215–225.

15 B. M. Tebo *et al.*, 'Microbial communities in dark oligotrophic volcanic ice cave ecosystems of Mt. Erebus, Antarctica', *Frontiers in Microbiology* 6, 179 (2015); H. Murtiyaningsih *et al.*, 'Selection of polymerase-producing thermophilic bacteria from Ijen Crater, East Java', *Indonesian Journal of Biotechnology and Biodiversity* 6, (2022): 26–32.

16 D. Cook *et al.*, 'Peering into the fire–An exploration of volcanic ecosystem services', *Ecosystem Services* 55 (2022): 101435.

17 As well as the lighter elements in our bodies, all the others are products of volcanism – some found in the rocks, such as calcium, sodium and phosphorous; others in the gases, including sulphur and chlorine. We don't eat volcanoes or their emissions, of course, but everything we tuck into derives ultimately from rock, water and gas. We simply rely on a complex food chain to turn the raw elements into something palatable and nutritious.

18 F. W. Junghuhn, '*Licht- en Schaduwbeelden uit de binnenlanden van Java*', (Amsterdam: Günst, 1867).

19 D. H. MacDonald *et al.*, 'Cougar Creek: quantitative assessment of obsidian use in the Greater Yellowstone ecosystem', *American Antiquity* 84, (2019): 158–178.

20 D. P. McGregor, 'Pele vs. geothermal: A clash of cultures', in *Bearing Dreams, Shaping Vision: Asia Pacific American Facing the 1990s* (Seattle: Washington State University Press, 1993): 45–60; A. Shih, 'The most perfect natural laboratory in the world: Making and knowing Hawaii National Park', *History of Science*

57 (2019): 493–517; R. Stoffle and K. Van Vlack, 'Talking with a volcano: Native American perspectives on the eruption of Sunset Crater, Arizona', *Land* 11, no. 2, 196 (2022).

21 P. Migoń and E. Pijet-Migoń, 'The role of geodiversity and geoheritage in tourism and local development, in L. Kubalíková *et al.*, (eds) *Visages of Geodiversity and Geoheritage*, Geological Society, London, Special Publications 530, (2023): doi: 10.1144/SP530-2022-115.

22 F. A. Perret, 'The Living Earth: A presentation of volcanology as a new and greater science', *Caribbean Center for Volcano Study*, Publication No. 1, 1940.

23 Basalt is made up of around forty per cent oxygen by mass – but it is not so easy to unleash it from the rock.

24 H. Follmann and C. Brownson, 'Darwin's warm little pond revisited: from molecules to the origin of life', *Naturwissenschaften* 96, no. 11 (2009): 1,265–1,292.

25 S. L. Miller, 'Production of some organic compounds under possible primitive earth conditions', *Journal of the American Chemical Society* 77 (1955): 2,351–2,361. A. P. Johnson *et al.*, 'The Miller volcanic spark discharge experiment', *Science* 322 (2008): 404. A. Lazcano and J. L. Bada, 'The 1953 Stanley L. Miller experiment: fifty years of prebiotic organic chemistry', *Origins of Life and Evolution of the Biosphere* 33 (2003): 235–242. I. Fry, 'The origins of research into the origins of life,' *Endeavour* 30 (2006): 24–28. A. Lazcano, 'Historical development of origins research', *Cold Spring Harbor perspectives in biology* 2 (2010): a002089.

26 W. Hamilton, 'IV. An account of the late eruption of Mount Vesuvius', *Philosophical Transactions of the Royal Society of London* 85 (1795): 73–116. See also: C. Cimarelli and K. Genareau, 'A review of volcanic electrification of the atmosphere and volcanic lightning', *Journal of Volcanology and Geothermal Research* 422 (2022): 107449.

27 M. D. Brasier *et al.*, 'Pumice as a remarkable substrate for the origin of life', *Astrobiology* 11, (2011): 725–735.

28 Another possibility is that the precursors of life were delivered to geothermal pools by meteorites. Some space rocks contain amino acids (which make up proteins), lipid molecules (precursors of cell membranes) and even sugars. For a deeper dive into the alternative theories, see: N. Kitada and S. Maruyama, 'Origins of building blocks of life: A review', *Geoscience Frontiers* 9 (2018): 1,117–1,153.

29 W. Martin *et al.*, 'Hydrothermal vents and the origin of life', *Nature Reviews Microbiology* 6, no. 11 (2008): 805–814.

Science fiction-like images of the giant worms and crustaceans thriving around the hot vents have now become familiar. Outlandish as they are, they are just specialized species of more familiar organisms. Photosynthesis is impossible in such deep, stygian waters. Instead, the food chain is supported by bacteria that harvest chemical energy from sulphur. For fascinating insights into their discovery, see: N. Oreskes, 'A context of motivation: US Navy oceanographic research and the discovery of sea-floor hydrothermal vents', *Social Studies of Science* 33 (2003): 697–742.

30 S. R. Gislason and E. H. Oelkers, 'Carbon storage in basalt', *Science* 344 (2014): 373–374.

31 D. E. Clark *et al.*, 'CarbFix2: CO_2 and H_2S mineralization during 3.5 years of continuous injection into basaltic rocks at more than 250°C', *Geochimica et Cosmochimica Acta* 279 (2020): 45–66.

32 R. Shaw and S. Mukherjee, 'The development of carbon capture and storage (CCS) in India: A critical review', *Carbon Capture Science & Technology* (2022): 100036.

Index